INTRODUCTION TO CONTROL THEORY FOR ENGINEERS

ALLAN SENSICLE, B.SC., C.ENG., M.I.E.E.

Senior Lecturer in Electrical Engineering, Chulalongkorn University, Bangkok
formerly Senior Lecturer, Hendon College of Technology

London · BLACKIE & SON LIMITED · Glasgow

First published 1968

216.88869.7 Cloth-bound edition
216.88870.0 Paper-bound editon

Printed in Great Britain by
Robert Cunningham and Sons Ltd, Alva

PREFACE

THIS BOOK IS INTENDED TO PROVIDE A BASIC UNDERSTANDING OF CONTROL system theory for students on engineering degree courses, and other courses of a similar level, with a control specialization.

I have always felt that the most efficient way to impart information is to present it in the simplest possible way. This has been attempted in the text that follows. A topic may be treated a little unconventionally in order not to complicate an explanation. (The reader will, no doubt, soon find that there is a wealth of highly academic textbooks on control theory.)

I wish to thank the Institution of Electrical Engineers for their kind permission to use some of their examination examples. I am indebted to my colleagues and students who by association have contributed so much to this book, in particular I would like to thank Mr Colin Chalmers for his valuable mathematical advice. I am most grateful for the help of Miss Ann Drybrough-Smith in advising me on the form of the manuscript.

The book would never have been completed without the understanding, patience and efficient typing of my wife, Pat.

ALLAN SENSICLE

Hendon College of Technology
November 1966

PREFACE

THIS BOOK IS INTENDED TO PROVIDE A BASIC UNDERSTANDING OF CONTROL system theory for students on engineering degree courses, and other courses of a similar level, with a control specialization.

I have always felt that the most efficient way to impart information is to present it in the simplest possible way. This has been attempted in the text that follows. A topic may be treated a little unconventionally in order not to complicate an explanation. (The reader will, no doubt, soon find that there is a wealth of highly academic textbooks on control theory.)

I wish to thank the Institution of Electrical Engineers for their kind permission to use some of their examination examples. I am indebted to my colleagues and students who by association have contributed so much to this book, in particular I would like to thank Mr Colin Chalmers for his valuable mathematical advice. I am most grateful for the help of Miss Ann Drybrough-Smith in advising me on the form of the manuscript.

The book would never have been completed without the understanding, patience and efficient typing of my wife, Pat.

ALLAN SENSICLE

Hendon College of Technology
November 1968

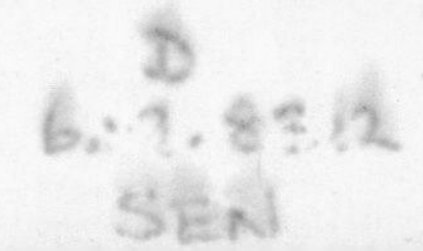

CONTENTS

CHAPTER 1

Introduction

1.1 How to use this textbook

The King of Hearts' advice 'begin at the beginning and go on till you come to the end: then stop' need by no means be applied to the reading of this textbook whose purpose is to explain some of the theory of control systems. Indeed a preliminary reading will take only an hour or so, during which time the student can achieve a simple but broad skeleton background of control systems; he may then gradually add flesh to the bones.

The background is obtained by reading in sequence all the sections marked with an asterisk. The other chapters and sections may then be digested as required. In general chapter sections are numbered. If, however, a capital letter is used instead of a number this indicates that the section is either standard mathematics or theory and practice applicable to other topics besides control systems.

Each illustration is directly integrated with the text, its explanation appearing below it.

The remote position control system is used throughout the book as an example, not because it is either the most common form or the most important, but because it is commonly found in educational establishments.

*1.2 What is an automatic control system?

It is unfortunate that we tend to complicate our already complicated technology by producing a 'language' of terms. These terms have to be explained and defined before embarking on the theory of the technology.

Automatic means self-acting, thus an automatic control system is a self-acting control system. Something which is self-acting could be a machine or a process, for example an electric motor is a self-acting device for producing a rotation, distillation is a self-acting process for the separation of liquids by vaporization and recondensation. Machines and processes are only automatic in as much as when power is supplied to them they act without human aid. Electricity is required by the electric motor and heat by the distillation process.

The behaviour of machines and processes may not be the same under different conditions. The speed of rotation of an electric motor may change according to the load the motor is turning, the rate of production of the end product in the distillation process depends on the temperature of the atmosphere surrounding the distillation apparatus.

We may assume that the speed of rotation of an electric motor can be controlled by the amount of electricity supplied to it and if the load being rotated is increased the speed falls. Suppose that the speed of rotation of an electric motor, driving a conveyor belt, is to be constant at 100 revolutions per minute. A human operator can control the speed of the electric motor by varying the supply of electricity to it, the speed of rotation being observed, by the operator, on a meter attached to the electric motor.

The procedure for controlling the speed of rotation so that it remains at 100 revolutions per minute is shown in figures 1.1, 1.2 and 1.3.

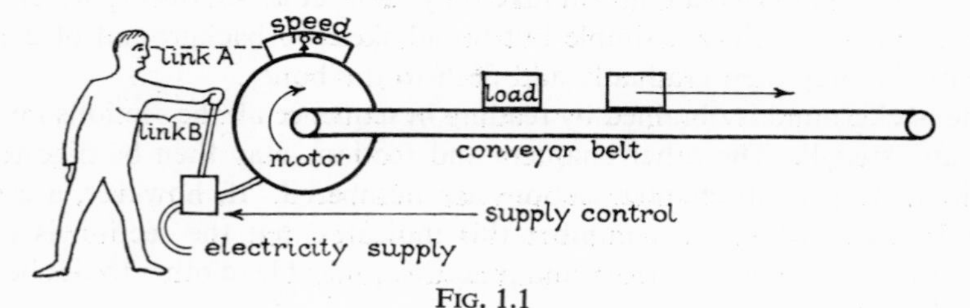

FIG. 1.1

Figure 1.1. Human operator observes that the speed is correct so does not change the amount of electricity supplied to the motor.

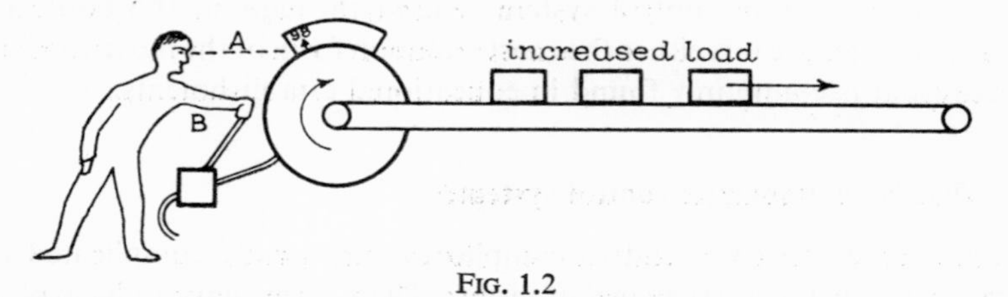

FIG. 1.2

Figure 1.2. Human operator observes that with the increased load the motor speed falls from 100 revolutions per minute to 98 revolutions per minute. He therefore increases the electricity supplied to the motor until the speed is 100 revolutions per minute.

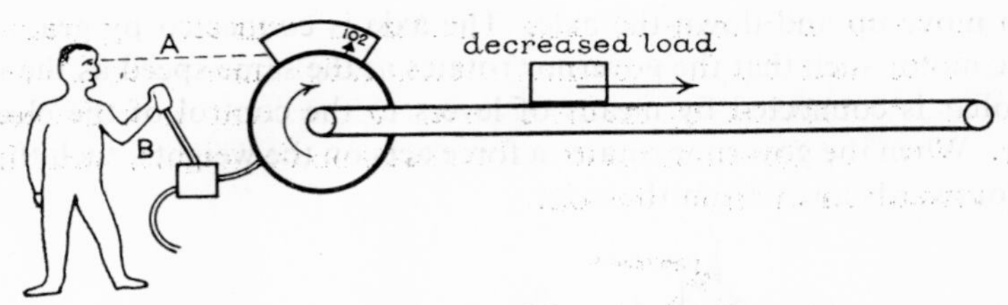

FIG. 1.3

Figure 1.3. Human operator observes that with the decreased load the motor speed increases from 100 revolutions per minute to 102 revolutions per minute. He therefore decreases the electricity supplied to the motor until the speed is 100 revolutions per minute.

The system that has been described is a speed control system, but as a human operator is involved it cannot be considered automatic. If we could replace the human operator by a device that automatically does his job, we would have an *automatic control system*.

Automatic control systems are sometimes called *servo-mechanisms*. Servo comes from the Latin *servus*, a slave, indicating that a servo-mechanism will slavishly do as required.

Automatic control systems are not new. James Watt was probably the first person to use an automatic speed control system. The Watt speed governor, shown in figure 1.4, could be used to control the speed of an electric motor automatically.

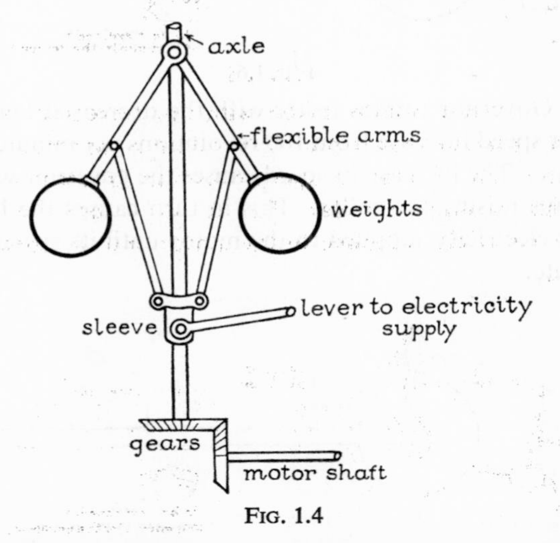

FIG. 1.4

Figure 1.4. The Watt speed governor consists of equal weights on flexible arms, attached at one end of the arms to an axle and at the other to a collar free to move up and down the axle. The axle is connected by gears to the electric motor such that the governor rotates at the same speed as the motor. The collar is connected by means of levers to the control of the electricity supply. When the governor rotates a force acts on the weights, such that they move outwards away from the axle.

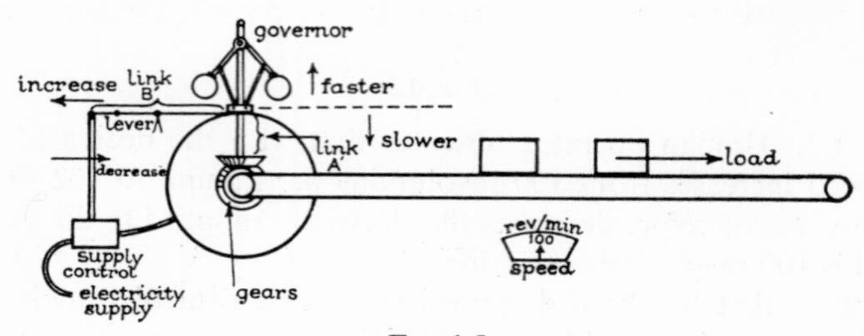

FIG. 1.5

Figure 1.5. With the correct speed of 100 revolutions per minute the weights of the governor are in such a position that the electricity supply control is not moved.

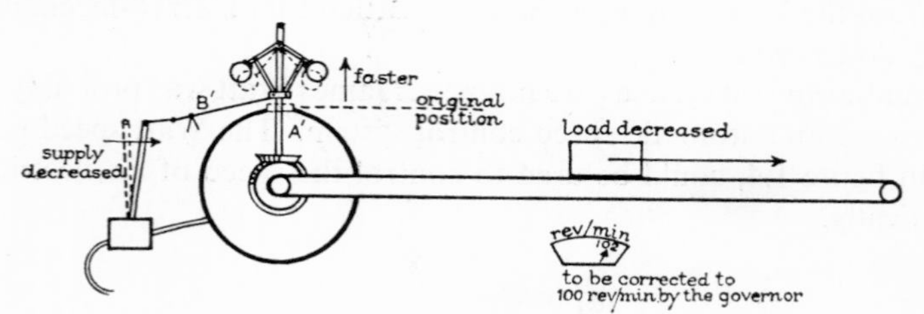

FIG. 1.6

Figure 1.6. Governor rotates faster with the decreased load. Motor and hence governor speed increase from 100 revolutions per minute to 102 revolutions per minute. The increase in speed causes the governor weights to move out further, thus raising the collar. This in turn causes the levers to move, decreasing the electricity supplied to the motor until its speed is 100 revolutions per minute.

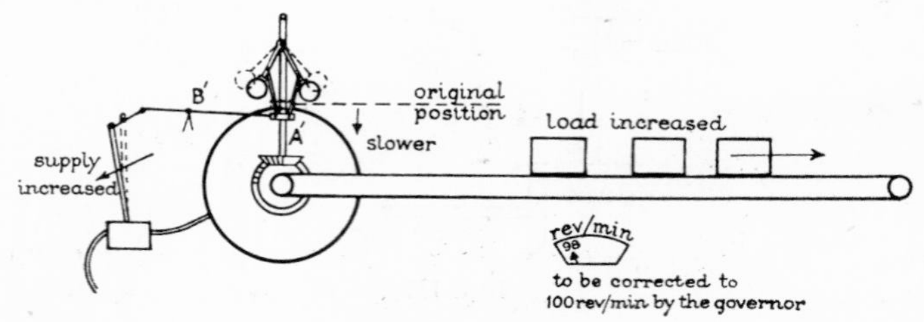

FIG. 1.7

Figure 1.7. Governor rotates slower with the increased load. Motor and hence governor speed decrease from 100 revolutions per minute to 98 revolutions per minute. The decrease in speed causes the governor weights to move in, thus lowering the collar. This in turn causes the levers to move, increasing the electricity supplied to the motor until its speed is 100 revolutions per minute.

The Watt speed governor thus replaces the function of the human operator completely. The visual link **A** in figure 1.1 is now the geared link **A**′ in figure 1.5. The arm controlling the supply, link **B** in figure 1.1, is now the lever link **B**′ in figure 1.5.

Hence a speed control system incorporating the Watt speed governor is an *automatic control system.*

Examining the automatic speed control system in figure 1.7 we see that it is connected mechanically and electrically in the form of a loop. It is redrawn more obviously in 'loop' form in figure 1.8.

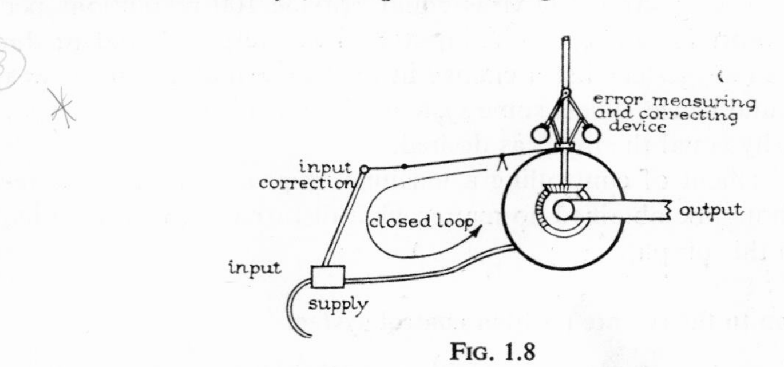

FIG. 1.8

Figure 1.8. The action of each element of the system has been specified regarding the part that it plays in controlling the system. Because the output and input of the system are joined in such a way that the output affects the input, the system is called a *closed-loop control system.* The significance of the closed loop will be explained thoroughly later in the text.

An example of an *open-loop control system* is shown in figure 1.9.

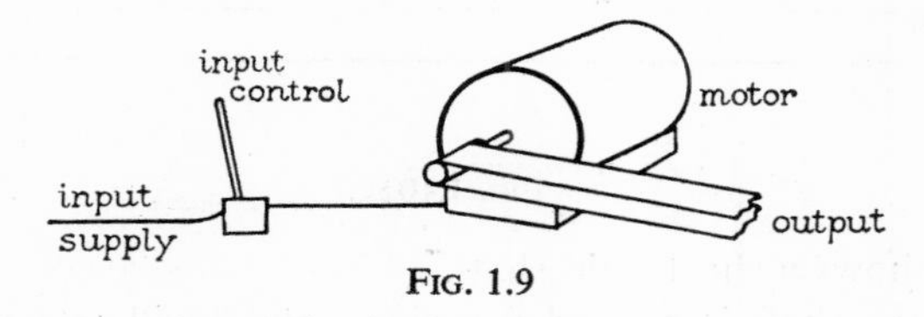

FIG. 1.9

Figure 1.9. This control system is the same as figure 1.1, but without the human operator. Open-loop control can only be used accurately for systems when there is no change in the output condition (i.e. in this case load), or for systems where accuracy is not important.

It is hoped that the specific examples given so far will give some idea of what control means, and we shall see shortly how they fit into the general theory and practice of control systems.

*1.3 The basic requirements of an automatic control system

The basic purpose of the automatic control of machines and processes is to obtain a predictable and consistent accuracy. If a control system is required, a closed-loop system will be far more effective than an open-loop system because it compares the output with the input of the system.

It should always be remembered that, for an ideal system or process, the *output* should be the same as the calibrated *input*. For example, if the speed of an electric motor is controlled by the supply of electricity to the motor, the rheostat or similar device giving the variation of supply will be calibrated not in volts but in speed (i.e. 100 volts could provide 100 revolutions per minute). This calibration indicates the input, and the output should be the same. It has been suggested that a change in load on a motor can cause a change in speed and hence, unless some system of control is used, the output will not necessarily equal the input as desired.

Another requirement of controlling a machine or process is that the required output should be obtained no matter what disturbance, within limits, is inflicted upon this output.

*1.4 Introduction to the remote position control system

We shall now introduce the remote position control system which will be used frequently as an example of an automatic control system.

Consider the shaft in the next figure.

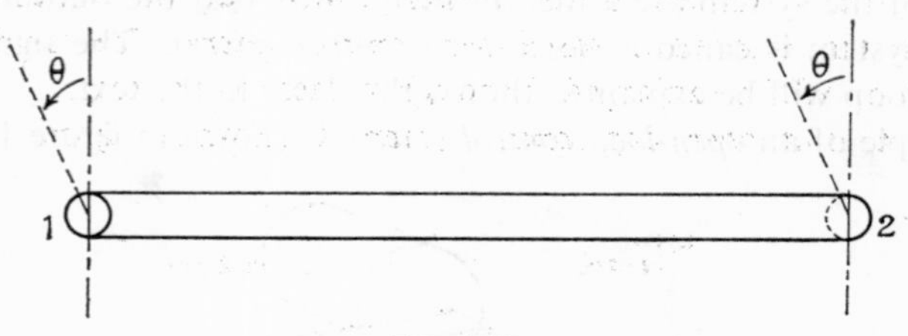

FIG. 1.10

Figure 1.10 shows a short rigid shaft.

If end 1 of the shaft is rotated θ radians, end 2 will rotate θ radians in exactly the same manner. This describes the function of a remote position control system. One end of the shaft is the input and the other end is the output, and a problem of control would exist if a load of, say, a ton had to be rotated at end 2 by a human operator at end 1.

Examples of this would be the positioning of the launching frame of a guided missile and the positioning of a radio telescope.

Obviously some aid is required by a human operator to position a large load. This can be provided electrically. Consider end 1, the input: the rotation of θ radians can be changed into an electrical quantity by means of a circular potentiometer. This is shown in the following figure.

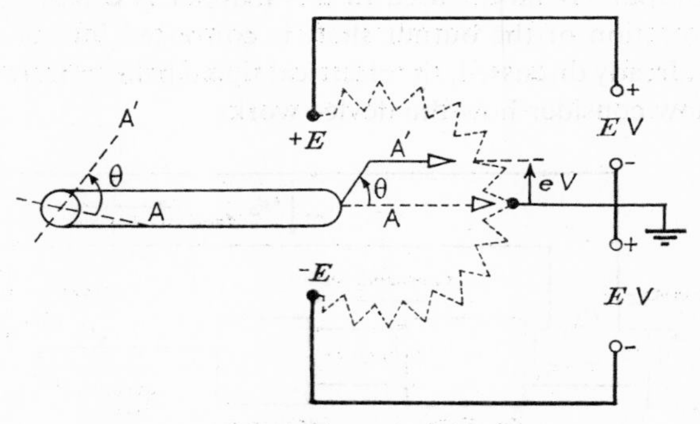

FIG. 1.11

Figure 1.11 shows the input shaft (end 1) attached to a circular potentiometer which is centre-tapped. The reference position A of the shaft is such that the wiper of the potentiometer is at earth potential. When the shaft is rotated anti-clockwise θ radians the wiper of the potentiometer also moves θ radians, such that the potential between earth and the wiper is e volts. Thus e volts represents θ radians of rotation. However, $+E$ volts and $-E$ volts are applied to each half of the potentiometer, therefore an anti-clockwise rotation is considered positive and a clockwise rotation negative. This is a means of showing a change in direction of rotation when considering potential.

The device is a *transducer*. (A transducer proportionally changes one form of energy into another, and in this case a rotation is changed into a potential.)

In further illustrations of this potentiometer the supplies for the potentials $+E$ volts and $-E$ volts will be omitted for simplicity as in the following figure.

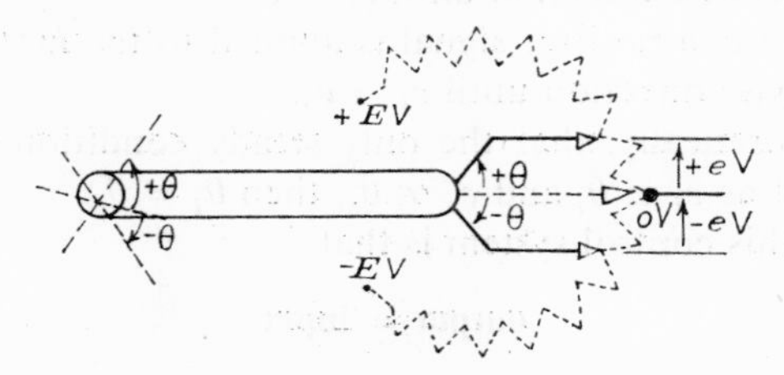

FIG. 1.12

Figure 1.12. In this figure it can be seen that

$$+\theta \text{ radians} \propto +e \text{ volts}$$
$$-\theta \text{ radians} \propto -e \text{ volts}$$

The signal of e volts can now be applied to an amplifier, the output of the amplifier being used as a supply for an electric motor, with the electric motor's shaft as the output. A motor used in this manner is usually called a *servo-motor*. The rotation of the output shaft is converted into electrical energy by the means already discussed, the electrical signal being referred to the input.

We shall now consider how the device works.

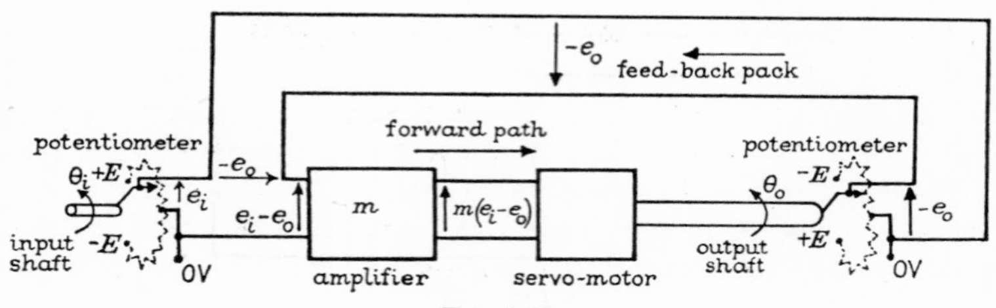

FIG. 1.13

Figure 1.13. When the input is at zero there is no electrical signal passing through the amplifier and none reaching the servo-motor, hence the system is at rest.

If the input is turned θ_i radians anti-clockwise, a signal of e_i volts ($\theta_i \propto e_i$) is applied to the amplifier and the output of the amplifier will be me_i volts, where m is the gain of the amplifier.

Now that the servo-motor has a supply it starts to rotate in the required direction.

Let the rotation of the output shaft be θ_o radians causing a signal of $-e_o$ volts to be fed back to the input ($\theta_o \propto -e_o$).

The signal at the amplifier is now $(e_i - e_o)$ volts. It follows that if $e_o = e_i$ the signal at the amplifier will be zero and the motor stops turning.

If e_o is greater than e_i a negative signal is applied to the motor and it rotates in the clockwise direction until $e_i = e_o$.

If e_o is less than e_i a positive signal is applied to the motor and it rotates in the anti-clockwise direction until $e_i = e_o$.

It follows, theoretically, that the only steady condition of the system is when $e_i = e_o$, but as $e_i \propto \theta_i$ and $e_o \propto \theta_o$, then $\theta_i = \theta_o$.

The object of this control system is that

$$\textbf{output} = \textbf{input}$$

and since $\theta_i = \theta_o$, in theory, this has been achieved.

We have seen that by using an amplifier a small input torque can position a very large load at the output.

Summary of operation

(*i*) The object of the system is that the output position shall be the same as the input position.

(*ii*) Input at zero. Output at zero.

(*iii*) Input set at θ_i radians causes a proportional electrical signal of e_i volts to activate the servo-motor.

(*iv*) Output position signal is fed back to input. This signal is negative and of value $-e_o$ volts at any time t seconds.

(*v*) New signal to activate the motor is $(e_i - e_o)$ volts.

(*vi*) If $e_i = e_o$ the system is at rest.

(*vii*) If $e_i < e_o$ the motor rotates in a clockwise direction until $e_i = e_o$.

(*viii*) If $e_i > e_o$ the motor rotates in anti-clockwise direction until $e_i = e_o$.

(*ix*) *Theoretically* $e_i = e_o$, therefore $\theta_i = \theta_o$.

We shall see later that this system is too crude to be useful, but it has enough of the basic elements of an automatic control system to be of use in developing a general theory.

*1.5 The block diagram and the transfer function

Most readers will be familiar with the block diagram, and in this book the control systems discussed will often be illustrated in the form of connected 'boxes' forming a block diagram. The boxes will not necessarily represent a complete device or even part of a device, but sometimes a function that occurs in the system. An example of the use of the block diagram is as follows:

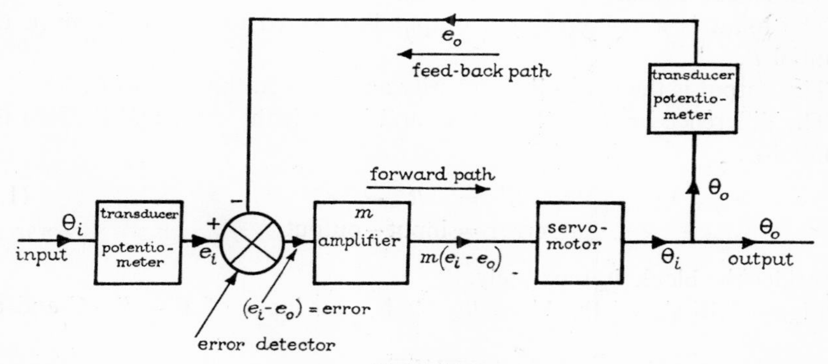

FIG. 1.14

Figure 1.14 shows the block diagram for the remote position control system discussed in the previous section. It is clearly seen from the diagram that this is a *closed-loop system*. The device which compares the output with the input is called an *error detector*, the error being the difference between the output and the input.

A block may be considered individually only when its output is unaffected by connection to the next block. In the example, figure 1.14, the gain of the amplifier should not be affected by connecting the servo-motor to its output, the output position of the servo-motor should not be affected by connecting the potentiometer to its shaft, etc.

We are now in a position to draw a general block diagram which will represent a basic closed-loop system, this is shown in the next figure.

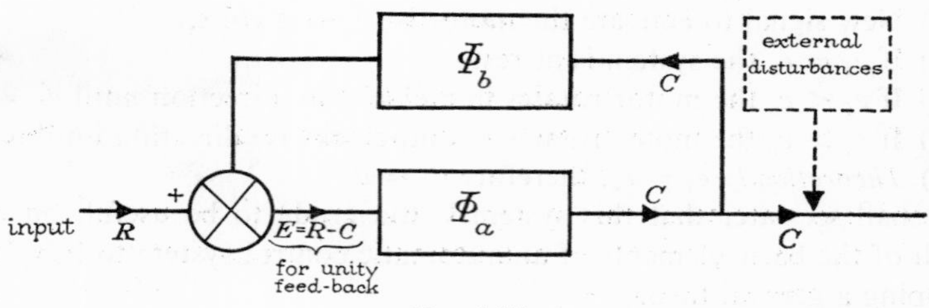

FIG. 1.15

Figure 1.15 shows the elements controlling the output lumped together as block Φ_a and in the feed-back loop there is another block Φ_b. In the remote position control system Φ_b is unity feed-back. (This loop can be considered to have unity feed-back because 'input' and 'output' transducers are identical, the input is e_i volts and the output is e_o volts.) Sometimes elements are required in the feed-back loop. A 'broken' block is used to represent external 'disturbances' causing a change in load.

The input is a reference (i.e. output is referred to it) and is given the symbol R.

The output is the controlled quantity and is given the symbol C.

The difference between the output and input is the error and is given the symbol E.

$$E = R - C \tag{1.1}$$
$$\text{error} = \text{input} - \text{output}$$

Consider the block Φ_a separately.

Figure 1.16 shows the block Φ_a. It has an input of $E = R - C$ and an

FIG. 1.16

output of C. The *transfer function* of this element, or group of elements, of the control system is defined as the operation of the element on its input to give an output, i.e.

$$\text{output } (C) = \underbrace{\text{transfer function}}_{\text{operation of the element/s}} \times \text{input } (E)$$

or

$$\frac{\text{output}}{\text{input}} = \text{transfer function}$$

The transfer function of the block will be called Φ_a.

$$\Phi_a = \frac{C}{E} \text{ where } C \text{ and } E \text{ are 'operational' functions} \quad (1.2)$$

To understand the transfer function a little better let us consider a simple electrical circuit.

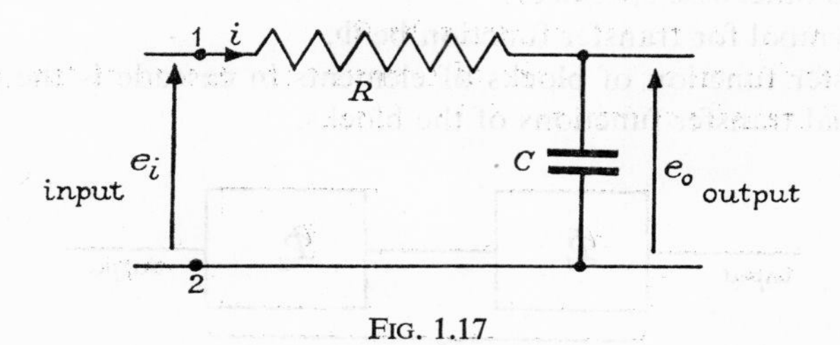

FIG. 1.17

Figure 1.17 shows a simple series circuit consisting of a resistance R and a capacitance C. An input voltage e_i is applied between terminals 1 and 2 and the output is the voltage across the capacitor C. We wish to find the transfer function of the circuit. From Kirchhoff's second law

$$e_i = iR + \frac{1}{C}\int_0^t i\,dt \quad (1.3)$$

where i is the current through the circuit at any time t, the capacitor being initially uncharged. However

$$\frac{1}{C}\int_0^t i\,dt = e_o$$

therefore

$$i = C\frac{de_o}{dt} \quad (1.4)$$

Substituting equation 1.4 in equation 1.3 we have

$$e_i = CR\frac{de_o}{dt} + e_o \quad (1.5)$$

Equation 1.5 is a first-order linear differential equation and d/dt can be replaced by an operator p. Equation 1.5 now becomes

$$e_i = CRpe_o + e_o \tag{1.6}$$

$$e_i = e_o(pCR+1)$$

$$\frac{e_o}{e_i} = \frac{\text{output}}{\text{input}} = \frac{1}{pCR+1} \tag{1.7}$$

Equation 1.7 is the transfer function of the circuit shown in figure 1.17. The input voltage e_i and the output voltage e_o now become functions of the operator (operational methods are dealt with more thoroughly in chapters 2 and 3).

A device is said to be *linear* if its transfer function is always the same irrespective of the system input. (All the devices we shall consider will be linear unless otherwise specified.)

Let the symbol for transfer function be Φ.

The transfer function of blocks of elements in cascade is the product of the individual transfer functions of the blocks.

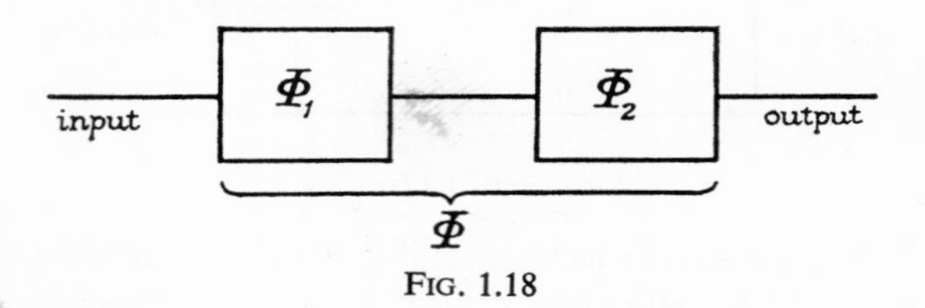

FIG. 1.18

Figure 1.18. This figure shows the two blocks of individual transfer functions Φ_1 and Φ_2 and, by definition, the overall transfer function Φ is

$$\Phi = \frac{\text{output}}{\text{input}}$$

$$\Phi = \underbrace{\frac{C'}{\text{input}}}\cdot\underbrace{\frac{\text{output}}{C'}}$$

$$\Phi = \Phi_1 \cdot \Phi_2 \tag{1.8}$$

It is very important to note that equation 1.8 is only true if the blocks do not interact with each other, for instance two circuits of the type in figure 1.17 interact when connected in cascade. We shall now find the transfer function Φ of the general closed-loop control system shown in figure 1.15. Firstly, let us consider a general single-loop system.

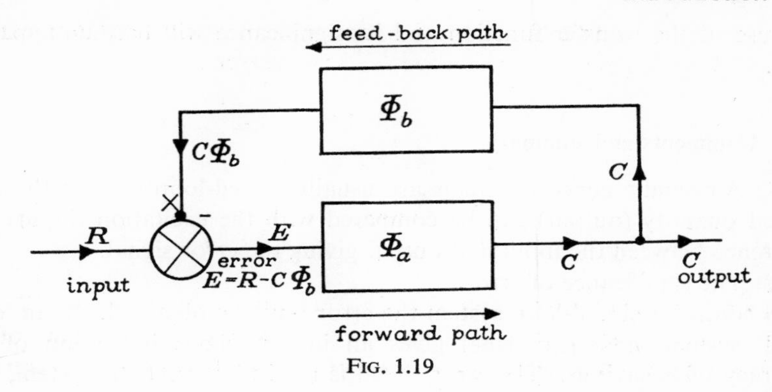

FIG. 1.19

Figure 1.19. Let the transfer function of the forward path be Φ_a and of the feed-back path be Φ_b. The closed-loop transfer function of the system Φ

$$\Phi = \frac{\text{output}}{\text{input}} = \frac{C}{R}$$

but

$$\Phi_a = \frac{C}{R - C\Phi_b}$$

$$C = \frac{\Phi_a R}{1+\Phi_a \Phi_b}$$

therefore

$$\frac{C}{R} = \Phi = \frac{\Phi_a}{1+\Phi_a\Phi_b} \tag{1.9}$$

Equation 1.9 is the closed-loop transfer function of the system shown in figure 1.19.

If a break is made in the feed-back path at X the output at the break divided by the system input is called the *open-loop transfer function.* It is often defined as the ratio of the feed-back signal $C\Phi_b$ to the actuating error signal E.

$$\frac{C\Phi_b}{E} = \text{open-loop transfer function} = \Phi_a\Phi_b$$

For the single-loop unity feed-back remote position control system $\Phi_b = 1$, the open-loop transfer function is Φ_a and the closed-loop transfer function is

$$\Phi = \frac{\Phi_a}{1+\Phi_a}$$

i.e. closed-loop transfer function $= \dfrac{\text{open-loop transfer function}}{1+\text{open-loop transfer function}}$

The use of the transfer function and its significance will become apparent later.

*1.6 Comments and summary

(*i*) Automatic control systems are usually closed-loop so that the controlled quantity (output) can be compared with the excitation (input), the difference between the input and output giving the error signal.

(*ii*) The significance of error.

To control a closed-loop system the error must be obtained. In an 'electrical' system the error signal gives an instantaneous indication of the accuracy of behaviour. The error signal is used to correct the system, in a closed-loop system it provides a control signal.

(*iii*) Open-loop control systems are only used where little or no change in output is expected, or for systems where accuracy is not important.

(*iv*) Block diagrams are used to give a clear representation of the elements making up a closed-loop system.

(*v*) The forward path of a system has a general transfer function Φ_a. The feed-back path of a general single-loop system has a transfer function Φ_b.

$$\text{closed-loop transfer function} = \frac{\Phi_a}{1+\Phi_a\Phi_b}$$

(*vi*) The remote position control system described in this chapter is not of practical use. We shall see this in the analysis of the system in chapter 4.

CHAPTER 2

The *p*-operator

The 'mathematics' we shall consider in this chapter is not intended as a simplified discussion of Laplace Transforms, but rather as an introduction to the use of operator methods in solving control system differential equations. The methods of solution of the differential equations discussed are valid, but the formal treatment of Laplace Transforms is to be found in chapter 3.

*2.1 An introduction to the *p*-operator

Consider the simple series circuit, as shown in figure 2.1, consisting of an inductance L, a resistance R and a battery of voltage E and of negligible internal resistance.

Figure 2.1. The switch is closed at a time $t=0$ seconds and it is required to find the output voltage e_o at any time t seconds.

The potential applied across terminals 1 and 2 is called a *step function* and is drawn as follows:

Figure 2.2 shows the graph of the step function which is described by

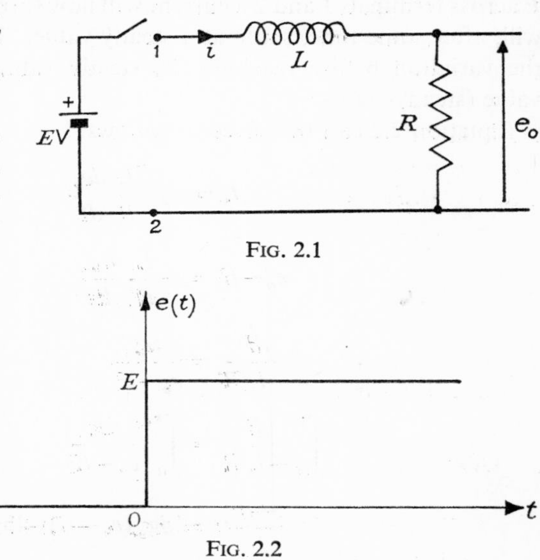

FIG. 2.1

FIG. 2.2

$$e(t) = 0 \text{ when } t < 0 \text{ and}$$

$$e(t) = E \text{ when } t > 0$$

where $e(t)$ represents a potential as a function of time.

Let i be the current flowing through the resistance-inductance combination at any time t.

Assuming the condition $i = 0$ when $t = 0$ and applying Kirchhoff's second law

$$E = iR + L\frac{di}{dt} \tag{2.1}$$

substituting $e_o = iR$ into equation 2.1 we have

$$E = e_o + \frac{L}{R}\frac{de_o}{dt} \tag{2.2}$$

This is a first-order linear differential equation. Like most linear differential equations where time is involved it has a composite solution consisting of a *transient part* and a *steady-state part.*

$$e_o = \underbrace{\text{steady-state solution} + \text{transient solution}}_{\text{complete solution}} \tag{2.3}$$

Both these solutions must be found to know e_o completely. After switching E across terminals 1 and 2 a current will flow through the circuit. This current will, after some time, settle at a steady value. The complete solution shows the variation before reaching the steady value (transient) and the steady value (steady-state).

Equation 2.2 can be solved as follows:

$$E = e_o + \frac{L}{R}\frac{de_o}{dt}$$

$$e_o - E = -\frac{L}{R}\frac{de_o}{dt}$$

$$\frac{dt}{-L/R} = \frac{de_o}{e_o - E}$$

$$\int_0^t \frac{dt}{-L/R} = \int_0^{e_o} \frac{de_o}{e_o - E}$$

$$\frac{-R}{L}t = \log_e (e_o - E) + \log_e K$$

where $\log_e K$ is a constant of integration. As $e_o = 0$ when $i = 0$ and $t = 0$, $K = -1/E$ and $-Ee^{-\frac{R}{L}t} = e_o - E$.

$$\text{Therefore} \qquad e_o = E - Ee^{-\frac{R}{L}t} \qquad (2.4)$$

$$e_o = \underbrace{\text{steady-state solution} + \text{transient solution}}_{\text{complete solution}}$$

We can see, by inspection, that E is the steady-state part and $-Ee^{-\frac{R}{L}t}$ is the transient part.

We shall now give symbols to the various operations performed. Let a step function be indicated by $1/p$, where $1/p$ is an operator. Hence in our example the input voltage as a function of time is the step function of magnitude E, applied to the circuit at time $t = 0$, and is now $E(1/p)$.

Let the operation of differentiation, with respect to time, be indicated by the operator p. ($1/p$ is taken to be the operation of integration with respect to time; the association of this operation with the step function should shortly become clear.)

Thus expressing equation 2.2 in terms of the p-operator we have

$$E = e_o + \frac{L}{R}\frac{d}{dt}e_o$$

(step function) (differentiation)

$$\frac{E}{p} = e_o + \frac{L}{R}p\,e_o$$

$$\frac{E}{p} = e_o\frac{L}{R}\left(\frac{R}{L}+p\right)$$

$$\text{therefore} \qquad e_o = E\frac{R}{L}\left(\frac{1}{p}\cdot\frac{1}{p+R/L}\right) \qquad (2.5)$$

We shall digress a little at this point and express the bracketed term of equation 2.5 as *partial fractions*. The partial fraction form of

$$\frac{1}{p}\cdot\frac{1}{p+R/L}$$

is

$$\frac{A}{p}+\frac{B}{p+R/L}$$

where A and B are constants. For these two expressions to be the same

$$\frac{1}{p}\cdot\frac{1}{p+R/L} = \frac{A}{p}+\frac{B}{p+R/L}$$

therefore

$$1 = A(p+R/L)+Bp$$

Equating coefficients of powers of p on each side, we have

$$0 = A+B \qquad \text{and} \qquad 1 = A\frac{R}{L}$$

giving $A = L/R$ and $B = -L/R$. Substituting these values of A and B into equation 2.5 we obtain

$$e_o = E\frac{R}{L}\left[\frac{L/R}{p}+\frac{-(L/R)}{p+R/L}\right]$$

$$e_o = E\frac{1}{p}-E\frac{1}{p+R/L} \tag{2.6}$$

To interpret the operator form we compare equation 2.6 with equation 2.4.

It seems reasonable that the function of the p-operator E/p is another way of describing the time function E, and that the function of the p-operator $E/(p+R/L)$ is another way of describing the time function $Ee^{-\frac{R}{L}t}$. Equations 2.4 and 2.6 are

$$e_o = E\,.\,1-E\,e^{-\frac{R}{L}t}$$

$$e_o = \underset{\text{steady-state part}}{E\,.\,\frac{1}{p}} \quad -E\,.\,\underset{\text{transient part}}{\frac{1}{p+R/L}}$$

By comparison we can say that the time function Ae^{-at} is equivalent to the p-operator function $A/(p+a)$ where A and a are constants. It should be noted that if a is zero the time function Ae^{-at} becomes the step function A and its equivalent p-operator function is A/p.

If equation 2.6 is expressed in a standard operator form we may readily obtain the solution as a function of time. Using the operational form of the step and exponential time functions we can now derive other relevant operational forms.

$$\frac{A}{p} \rightarrow A \text{ (function of time) where } A \text{ is constant} \tag{2.7}$$

$$\frac{1}{p+A} \rightarrow e^{-at} \text{ where } a \text{ is constant} \tag{2.8}$$

(An arrow is used to indicate the change from a function of the p-operator to a function of time and vice-versa.)

As

$$\sinh at = \frac{e^{+at} - e^{-at}}{2}$$

$$\sinh at \rightarrow \tfrac{1}{2}\left(\frac{1}{p-a} - \frac{1}{p+a}\right)$$

$$\sinh at \rightarrow \frac{a}{p^2 - a^2} \qquad (2.9)$$

Similarly

$$\cosh at \rightarrow \frac{p}{p^2 - a^2} \qquad (2.10)$$

Similarly, as

$$\sin at = \frac{e^{+jat} - e^{-jat}}{2j}$$

where j is the complex operator $\sqrt{-1}$,

$$\sin at \rightarrow \frac{a}{p^2 + a^2} \qquad (2.11)$$

and

$$\cos at \rightarrow \frac{p}{p^2 + a^2} \qquad (2.12)$$

In further equations in the text functions of the p-operator will be given capital letters to distinguish them from functions of time t, which will have small letters (e.g. e_o is a function of time t, whereas E_o is a function of the operator p).

Often only the steady-state solution of a linear differential equation is required. It is obtained quite easily from the operational form of the differential equation. This is illustrated by taking equation 2.6 and rearranging it in its transfer function form.

$$\frac{\text{output}}{\text{input}} = \frac{E_o}{E/p} = \frac{R/L}{p + R/L} \qquad (2.13)$$

For the input to be a step function the steady-state condition must be such that there is no rate of change on the output, i.e. terms involving d/dt will be zero, therefore $p = 0$.

Thus if $p = 0$ in equation 2.13 the output will be the steady-state output.

$$\frac{E_{o\ \text{steady-state}}}{E/p} \rightarrow 1$$

$$\therefore \quad E_{o\ \text{steady-state}} \rightarrow \frac{E}{p} \tag{2.14}$$

The steady-state output in operational form, equation 2.14 is the same as previously obtained in equation 2.6. We shall see that the transient part of the equation is obtained by subtracting the steady-state part E/p from equation 2.6 (the complete solution).

The assumption is made that $p \rightarrow d/dt$, $p^2 \rightarrow d^2/dt^2$, etc.

Mathematicians, quite rightly, may object to these assumptions, but as has already been stated this section is intended to introduce the idea of operator methods for solving linear differential equations.

*2A Partial fractions

A transfer function can be expressed as the ratio of two polynomials in terms of the p-operator. Let a general transfer function be Φ,

$$\Phi = \frac{a_0 + a_1 p + a_2 p^2 + a_3 p^3 + \ldots + a_n p^n}{b_0 + b_1 p + b_2 p^2 + b_3 p^3 + \ldots + b_m p^m} \tag{2.15}$$

where $a_0, a_1, a_2, \ldots a_n$ and $b_0, b_1, b_2, \ldots b_m$ are constant coefficients and n is generally less than m. To make the theory a little less complicated we shall consider a specific Φ. As an example let

$$\Phi = \frac{2p^2 + 3p + 2}{(p^2+4)(p+4)(p+1)^2} \tag{2.16}$$

We can express equation 2.16 in partial fractions as long as:

(*i*) the linear factor $p+4$ of the denominator has the partial fraction of $A_1/(p+4)$ where A_1 is a constant,

(*ii*) the quadratic factor p^2+4 of the denominator has the partial fraction of $(A_2 p + A_3)/(p^2+4)$ where A_2 and A_3 are constants,

(*iii*) the squared linear factor $(p+1)^2$ of the denominator has the partial fractions

$$\frac{A_4}{p+1} + \frac{A_5}{(p+1)^2}$$

where A_4 and A_5 are constants.

The transfer function Φ can now be expressed as partial fractions as follows,

$$\Phi = \frac{A_1}{p+4}+\frac{A_2p+A_3}{p^2+4}+\frac{A_4}{p+1}+\frac{A_5}{(p+1)^2} \tag{2.17}$$

Equating the right-hand sides of equations 2.17 and 2.16 we obtain

$$\frac{A_1}{p+4}+\frac{A_2p+A_3}{p^2+4}+\frac{A_4}{p+1}+\frac{A_5}{(p+1)^2} = \frac{2p^2+3p+2}{(p^2+4)(p+4)(p+1)^2}$$

therefore $A_1(p^2+4)(p+1)^2+(A_2p+A_3)(p+4)(p+1)^2+$

$$A_4(p^2+4)(p+4)(p+1)+A_5(p^2+4)(p+4) = 2p^2+3p+2 \tag{2.18}$$

Equation 2.18 must be true for all values of p. Hence $p^2 = -4$ (i.e. $p = \pm\sqrt{-4} = \pm j2$), $p = -4$ and $p = -1$, respectively make the factors p^2+4, $p+4$ and $(p+1)^2$ equal to zero.
With $p = +j2$ we obtain

$$(A_2 j2+A_3)(-20+j10) = -6+j6 \tag{2.19}$$

With $p = -j2$ we obtain

$$[A_2(-j2)+A_3](-20-j10) = -6-j6 \tag{2.20}$$

Equations 2.19 and 2.20 give

$$A_2 = \tfrac{-3}{50} \quad \text{and} \quad A_3 = \tfrac{9}{25}$$

With $p = -4$ we obtain $\quad A_1 = \frac{11}{90}$

With $p = -1$ we obtain $\quad A_5 = \frac{1}{15}$

To find A_4 we must equate the coefficients of p on each side. It is not important which power is chosen as long as A_4 appears in the resulting equation. Choosing coefficients of p^4 we have

$$A_1+A_2+A_4 = 0 \tag{2.21}$$

Substituting the values of A_1 and A_2 in equation 2.21 we have

$$A_4 = -\tfrac{14}{225}$$

We have obtained the coefficients of the partial fractions. The transfer function equation in partial-fraction form is

$$\Phi = \frac{11/90}{p+4}+\frac{-3/50p+9/25}{p^2+4}+\frac{-14/225}{p+1}+\frac{1/15}{(p+1)^2} \tag{2.22}$$

Partial fractions enable a complicated transfer function, in operator form, to be expressed in a form which makes a 'translation' into a solution as a function of time easier.

*2.2 Further operation notations

Taking the 'equation' $\sin at \to a/(p^2+a^2)$ ('equation' 2.11) let us find the effect on $\sin at$ of changing the p^2 term to $(p+b)^2$ where b is a constant. The right-hand side of the equation becomes

$$\frac{a}{(p+b)^2+a^2}$$

This can be simplified by means of partial fractions to give

$$\frac{j/2}{(p+b)+ja}+\frac{-j/2}{(p+b)-ja}$$

where $(p+b)+ja$ and $(p+b)-ja$ are the factors of $(p+b)^2+a^2$. The partial-fraction expression can be translated into a function of time by comparing it with 'equation' 2.8 $[1/(p+a) \to e^{-at}]$.

$$\frac{j/2}{p+(b+ja)}-\frac{j/2}{p+(b-ja)} \to \frac{j}{2}(e^{-(b+ja)t}-e^{-(b-ja)t})$$

$$\to \frac{e^{-bt}}{j2}(e^{jat}-e^{-jat})$$

$$\therefore \quad \frac{a}{(p+b)^2+a^2} \to e^{-bt}\sin at \qquad (2.23)$$

We can conclude that $e^{-bt}\sin at$ has the operational notation

$$\frac{a}{(p+b)^2+a^2}$$

Similarly it can be shown that

$$\frac{p+b}{(p+b)^2+a^2} \to e^{-bt}\cos at \qquad (2.24)$$

$$\frac{a}{(p+b)^2-a^2} \to e^{-bt}\sinh at \qquad (2.25)$$

$$\frac{p+b}{(p+b)^2-a^2} \to e^{-bt}\cosh at \qquad (2.26)$$

Let us examine what happens if we divide 'equation' 2.23 by a and then we let a tend to zero.

$$\left[\frac{1}{(p+b)^2+a^2} \rightarrow \frac{e^{-bt}\sin at}{a}\right]_{a\rightarrow 0}$$

When a tends to zero sin at becomes at, therefore

$$\frac{1}{(p+b)^2} \rightarrow e^{-bt}\frac{at}{a}$$

$$\frac{1}{(p+b)^2} \rightarrow e^{-bt}\,t \tag{2.27}$$

We are now in a position to draw up a table of the operational notations we have obtained. The notations consist of 'equations' 2.7 to 2.12 and 2.23 to 2.27.

Table of Operational Notation

Operational notation (Function of p)	Translation (Function of t)	'Equation' number in the text
$\frac{A}{p}$ where A is constant	A where A is a constant function of time	2.7
$\frac{1}{p+a}$	e^{-at}	2.8
$\frac{a}{p^2-a^2}$	sinh at	2.9
$\frac{p}{p^2-a^2}$	cosh at	2.10
$\frac{a}{p^2+a^2}$	sin at	2.11
$\frac{p}{p^2+a^2}$	cos at	2.12
$\frac{a}{(p+b)^2+a^2}$	e^{-bt} sin at	2.23
$\frac{p+b}{(p+b)^2+a^2}$	e^{-bt} cos at	2.24
$\frac{a}{(p+b)^2-a^2}$	e^{-bt} sinh at	2.25
$\frac{p+b}{(p+b)^2-a^2}$	e^{-bt} cosh at	2.26
$\frac{1}{(p+b)^2}$	te^{-bt}	2.27

*2.3 Example

Solution of a second-order linear differential equation

Let the input to a system be y, where y is a function of time t. The input is 'operated on' by the system to give an output of x, say, where x is a function of time t.

Let the system be described by the linear second-order differential equation

$$y = a\frac{d^2x}{dt^2}+b\frac{dx}{dt}+cx \tag{2.28}$$

where a, b and c are constants. The input and the output of the system described as functions of the p-operator will be Y and X respectively.

Putting equation 2.28 into the operational form we have

$$Y = ap^2X+bpX+cX \tag{2.29}$$

Let all terms of equation 2.29 be zero at time $t = 0$. The solution of equation 2.29 will be in two parts, the steady-state part and the transient part.

$$x = x_{\text{steady-state}}+x_{\text{transient}} \qquad \text{function of time } t$$

$$X = X_{\text{steady-state}}+X_{\text{transient}} \qquad \text{function of operator } p$$

Let the input be a step function of magnitude A. To find the steady-state solution with a step function input, $Y = A/p$, we make p zero in equation 2.29 (this has been discussed in section *2.1).

$$Y = X_{\text{steady-state}}[ap^2+bp+c]_{p=0}$$

$$\therefore \quad X_{\text{steady-state}} = \frac{Y}{c} = \frac{A}{pc}$$

To find the transient solution we subtract the steady-state solution from the complete solution,

$$X_{\text{transient}} = X_{\text{complete solution}} - X_{\text{steady-state}}$$

$$X_{\text{transient}} = \frac{Y}{ap^2+bp+c}-\frac{Y}{c}$$

$$X_{\text{transient}} = \left[\frac{A/p}{ap^2+bp+c}-\frac{A}{pc}\right] \tag{2.30}$$

The first expression on the right-hand side of equation 2.30 can give three solutions depending upon the relationship between a, b and c. The quadratic in equation 2.30 can be rearranged, by completing the square

$$a\left[p^2+\frac{b}{a}p+\frac{c}{a}\right] = a\left[\left(p+\frac{b}{2a}\right)^2+\left(\frac{c}{a}-\frac{b^2}{4a^2}\right)\right]$$

The three solutions will depend on whether c/a is greater than, less than, or equal to $b^2/4a^2$.

Transient solution (*i*) $c/a > b^2/4a^2$

$$X_{\text{transient}} = \frac{A}{a}\cdot\frac{1}{p}\left(\frac{1}{(p+b/2a)^2+(c/a-b^2/4a^2)}\right)-\frac{A}{pc} \tag{2.31}$$

Let $(c/a-b^2/4a^2) = \beta^2$. Equation 2.31 can be expressed in the following partial fractions (section 2A, p. 20):

$$X_{\text{transient}} = \frac{A}{a}\left(\frac{A_1}{p}+\frac{A_2p+A_3}{(p+b/2a)^2+\beta^2}\right)-\frac{A}{pc}$$

where $A_1 = a/c$, $A_2 = -a/c$ and $A_3 = -b/c$.
Therefore

$$X_{\text{transient}} = \frac{A}{pc}-\frac{A}{a}\left(\frac{pa/c+b/c}{(p+b/2a)^2+\beta^2}\right)-\frac{A}{pc}$$

$$X_{\text{transient}} = -\frac{A}{c}\left(\frac{p+b/2a}{(p+b/2a)^2+\beta^2}+\frac{b/2a}{(p+b/2a)^2+\beta^2}\right) \tag{2.32}$$

Equation 2.32 can be expressed as a function of time using 'equations' 2.23 and 2.24.

$$x_{\text{transient}} = -\frac{Ae^{-\frac{b}{2a}t}}{c}\left(\cos\beta t+\frac{b}{2a\beta}\sin\beta t\right) \tag{2.33}$$

There is, however, a better way of expressing the transient solution. Consider the expression

$$x_{\text{transient}} = -\frac{A}{c\beta}\sqrt{\left(\frac{c}{a}\right)}\,e^{-\frac{b}{2a}t}\left(\sqrt{\left(\frac{a}{c}\right)}\,\beta\cos\beta t+\frac{b}{2\sqrt{(ac)}}\sin\beta t\right) \tag{2.34}$$

This equation can be expressed in the form

$$-\frac{A}{c\beta}\sqrt{\left(\frac{c}{a}\right)}\,e^{-\frac{b}{2a}t}(\cos\phi\sin\beta t+\sin\phi\cos\beta t) =$$

$$= -\frac{A}{c\beta}\sqrt{\left(\frac{c}{a}\right)}\,e^{-\frac{b}{2a}t}\sin(\beta t+\phi) \tag{2.35}$$

If we compare the expression in the bracket in equation 2.34 with the expression in the bracket in equation 2.35 we have

$$\cos\phi = \frac{b}{2\sqrt{(ac)}} \quad \text{and} \quad \sin\phi = \sqrt{\left(\frac{a}{c}\right)}\beta$$

This gives

$$\tan\phi = \frac{2a\beta}{b}$$

Providing we specify

$$\phi = \tan^{-1}\frac{2a\beta}{b} \quad \text{and} \quad \beta = \sqrt{\left(\frac{c}{a}-\frac{b^2}{4a^2}\right)}$$

the transient solution can be written

$$x_{\text{transient}} = -\frac{A}{c\beta}\sqrt{\left(\frac{c}{a}\right)}\,e^{-\frac{b}{2a}t}\sin(\beta t+\phi) \tag{2.36}$$

Transient solution (ii) $c/a < b^2/4a^2$

The solution for this condition is obtained in exactly the same way as for solution (*i*), equation 2.36, and the result is of a similar form.

$$X_{\text{transient}} = -\frac{A}{c}\left[\frac{p+b/2a}{(p+b/2a)^2-\beta^2}-\frac{b/2a}{(p+b/2a)^2-\beta^2}\right]$$

β is now taken as $b^2/4a^2 - c/a$ to indicate clearly that $c/a < b^2/4a^2$. Thus we see that the transient solution will be

$$x_{\text{transient}} = -\frac{A}{c\beta}\sqrt{\left(\frac{c}{a}\right)}\,e^{-\frac{b}{2a}t}\sinh(\beta t+\phi) \tag{2.37}$$

where

$$\phi = \tanh^{-1}\frac{2a\beta}{b} \quad \text{and} \quad \beta = \sqrt{\left(\frac{b^2}{4a^2}-\frac{c}{a}\right)}$$

Transient solution (iii) $c/a = b^2/4a^2$

The solution for this condition is obtained in the same way as solutions (*i*) and (*ii*), equations 2.36 and 2.37. The result is of the form

$$x_{\text{transient}} = -\frac{A}{c}\,e^{-\frac{b}{2a}t}\left(1+\frac{b}{2a}t\right) \tag{2.38}$$

Summary

The solutions of a second-order linear differential equation depend on the value of the equation's coefficients. The three *complete* solutions for the equation $Y = (ap^2+bp+c)X$, with the input Y a step function of magnitude A, are given for the various conditions as follows:

(*i*) $c/a > b^2/4a^2$

$$x = \frac{A}{c}\left[1-\frac{1}{\beta}\sqrt{\left(\frac{c}{a}\right)}\,e^{-\frac{b}{2a}t}\sin(\beta t+\phi)\right] \tag{2.39}$$

where $\phi = \tan^{-1}\dfrac{2a\beta}{b}$ and $\beta = \sqrt{(c/a-b^2/4a^2)}$.

(*ii*) $c/a < b^2/4a^2$

$$x = \frac{A}{c}\left[1-\frac{1}{\beta}\sqrt{\left(\frac{c}{a}\right)}\,e^{-\frac{b}{2a}t}\sinh(\beta t+\phi)\right] \tag{2.40}$$

where $\phi = \tanh^{-1}\dfrac{2a\beta}{b}$ and $\beta = \sqrt{(b^2/4a^2-c/a)}$.

(*iii*) $c/a = b^2/4a^2$

$$x = \frac{A}{c}\left[1-e^{-\frac{b}{2a}t}\left(1+\frac{b}{2a}t\right)\right] \tag{2.41}$$

where $\beta = 0$.

The conditions for the three solutions are the same conditions for the roots of the quadratic ap^2+bp+c to be complex, real and equal. The roots of ap^2+bp+c are given by

$$-\frac{b}{2a}\pm\sqrt{(b^2/4a^2-c/a)}$$

for complex roots	(condition (*i*))	$c/a > b^2/4a^2$
for real roots	(condition (*ii*))	$c/a < b^2/4a^2$
for equal roots	(condition (*iii*))	$c/a = b^2/4a$

It is important to note that the equations considered all have zero initial conditions (this is usually the case for linear control system equations). Linear differential system equations with non-zero initial conditions are considered in the formal mathematics in chapter 3.

*2.4 The impulse function

A transfer function is the ratio of output to input as functions of an operator.

$$\frac{\text{output (function of } p)}{\text{input (function of } p)} = \text{transfer function (function of } p)$$

The output (function of p) will give the transfer function (function p) if the input to a system, as a function of the p-operator, is unity. The unit input is called an *impulse function.* This form of input is extremely useful for system analysis and is dealt with more thoroughly in chapter 3. The solution resulting from a system equation, with an impulse function input, as a function of time is called the *impulse response.* The impulse response can be said to identify a system.

2.5 Examples

1. Find an expression for the variation of x with time t if

$$y = x+10\frac{dx}{dt}$$

where the function of time y is a step function of magnitude 5, starting at time $t = 0$, and the initial conditions are zero.

$$[x = 5(1-e^{-0\cdot 1t})]$$

2. Find an expression for the variation of θ_o with time t if

$$\theta_i = 2\theta_o+2\frac{d\theta_o}{dt}$$

where the function of time θ_i is a step function of magnitude 2, starting at time $t = 0$, and the initial conditions are zero.

$$(\theta_o = 1-e^{-t})$$

3. A system has an input R and an output C; the transfer function of the system is Φ

$$\Phi = \frac{1}{p0\cdot 5+1}$$

Find the variation of the output with time if the input is a unit impulse function at time $t = 0$.

$$(C = 2e^{-2t})$$

4. An electric circuit consists of an initially uncharged 50 μF capacitor connected in series with a 500 Ω resistor. A direct voltage of 100 V is suddenly applied to the circuit. Find expressions for the subsequent variation of voltage across the capacitor and current through the circuit.

$$[100(1-e^{-40t});\ 0\cdot 2\,e^{-40t}]$$

5. Find an expression for the variation of x with time t if

$$y = \frac{d^2x}{dt^2}+4\frac{dx}{dt}+6x$$

where the function of time y is a step function of magnitude 60 starting at time $t = 0$, and the initial conditions are zero.

$$\{x = 10\,[1-1{\cdot}732\,e^{-2t}\sin(1{\cdot}414t+0{\cdot}615)]\}$$

6. A control system is described by the transfer function

$$\frac{C}{R} = \frac{10}{p^2+2p+1}$$

where C and R are functions of the operator p. Find an expression for the system output as a function of time t if the control system input is a unit step function starting at time $t = 0$. Sketch the variation of the output with time.

$$\{10\,[1-(1+t)\,e^{-t}]\}$$

7. A remote position control system has the differential equation

$$\omega_n^2\theta_i = \frac{d^2\theta_o}{dt^2}+2\zeta\omega_n\frac{d\theta_o}{dt}+\omega_n^2\theta_o$$

Find an expression for the output position θ_o as a function of time t given the following information,

input, step function,	$\theta_i = 1$ radian,
undamped natural frequency,	$\omega_n = 1$ radian per second,
damping ratio,	$\zeta = 3{\cdot}75$.

$$[1-0{\cdot}277\,e^{-3{\cdot}75t}\sinh(3{\cdot}614t+1{\cdot}994)]$$

8. A control system has an open-loop transfer function

$$\frac{K}{p(p+2)(p+3)}$$

where K is the gain constant. Find the system impulse response if K is unity.

$$(\tfrac{1}{6}-\tfrac{1}{2}\,e^{-2t}+\tfrac{1}{3}\,e^{-3t})$$

CHAPTER 3

The Laplace Transform

3A Introduction

It is often convenient, for the purposes of computation, to express a problem in terms of a variable that is different from the one in which it was originally stated. For example, to obtain the quotient of two complex numbers given in the $a+jb$ form it is easier to express them in the polar form. Another example is the use of logarithms to simplify calculations. Suppose we introduce the complex variable $s = \alpha + j\omega$ where α and ω are quantities having the dimensions of 1/time. We can then find a relationship which will transfer a given function of time $f(t)$ into a function of the complex variable $\bar{f}(s)$, so that there is a unique $\bar{f}(s)$ for all $f(t)$. The transformation of $f(t)$ into $\bar{f}(s)$, where s is called the *Laplace variable*, is defined by the integral

$$\bar{f}(s) = \int_{0-}^{\infty} f(t)\, \mathrm{e}^{-st}\, dt \tag{3.1}$$

$\bar{f}(s)$ is called the *Laplace transform* of $f(t)$.

To ensure that time functions which occur only at $t = 0$ (the unit impulse function—section 3B, number (*vi*)) are included in the defining integral, the lower limit of integration is $t = 0-$. We will use a bar sign to indicate a function of s. All other variables will be functions of time t unless otherwise specified. Thus, for example, $\bar{x}$ is the Laplace transform of x. Equation 3.1 is usually written

$$\bar{f}(s) = \mathscr{L} f(t)$$

where $\mathscr{L}$ means 'Laplace transform of'.

We shall now determine the Laplace transforms of functions and operations and investigate their use in the solution of differential equations.

It is important to note, from the integral defining the Laplace transform (equation 3.1), that the time function is considered only within the limits $t = 0-$ to $t = \infty$ and not before $t = 0-$.

3B The Laplace transforms of functions and operations

Functions

(*i*) *The step function*

A step function is defined as follows

$$f(t) = 0 \quad \text{when} \quad t < 0$$

$$f(t) = A \quad \text{when} \quad t > 0,$$

where A is constant.

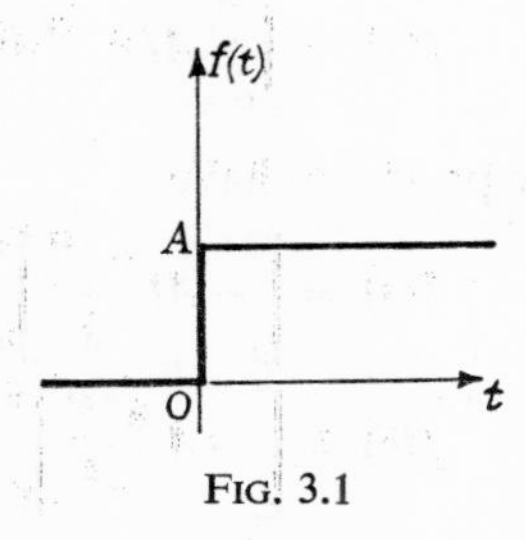

FIG. 3.1

Figure 3.1 shows the step function under consideration. Using equation 3.1 and the definition of the step function from time $t = 0$ we can find its Laplace transform.

$$f(t) = A$$

$$f(s) = \int_{0-}^{\infty} A\,\mathrm{e}^{-st}\,dt$$

$$f(s) = \left[-A\frac{\mathrm{e}^{-st}}{s}\right]_{0-}^{\infty}$$

$$f(s) = A/s \tag{3.2}$$

(*ii*) *The ramp function*

A ramp function is defined as follows

$$f(t) = 0 \quad \text{when} \quad t < 0$$

$$f(t) = At \quad \text{when} \quad t > 0,$$

where A is constant.

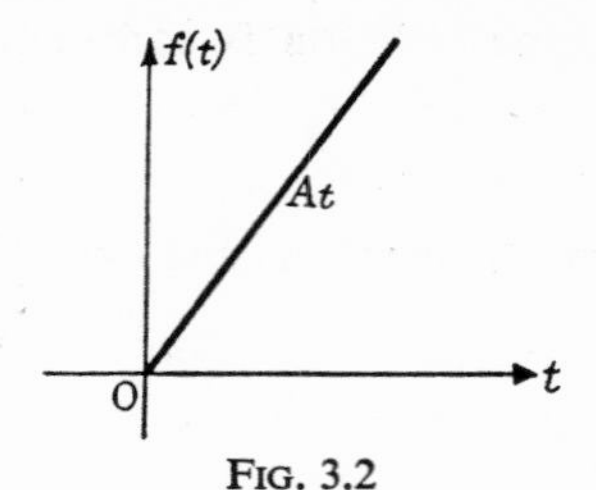

FIG. 3.2

Figure 3.2 shows the ramp function under consideration. Using equation 3.1 and the definition of the ramp function, from time $t = 0$, we can find its Laplace transform.

$$f(t) = At$$

$$f(s) = \int_{0-}^{\infty} At\, e^{-st}\, dt \tag{3.3}$$

Integrating equation 3.3 by parts, we have

$$f(s) = \left[-At\frac{e^{-st}}{s}\right]_{0-}^{\infty} - \int_{0-}^{\infty} -A\frac{e^{-st}}{s}\,dt$$

$$f(s) = \left[-A\frac{e^{-st}}{s^2}\right]_{0-}^{\infty}$$

$$f(s) = \frac{A}{s^2} \tag{3.4}$$

and it can be shown that

$$\mathscr{L}At^n = \frac{An!}{s^{n+1}} \tag{3.5}$$

where A is constant and n is a positive integer.

(iii) Exponential functions

Consider the exponential function e^{-at}, where a is constant. Using equation 3.1 we can find its Laplace transform

$$f(t) = e^{-at}$$

$$f(s) = \int_{0-}^{\infty} e^{-at} e^{-st}\, dt$$

$$f(s) = \left[\frac{e^{-(s+a)t}}{-(s+a)}\right]_{0-}^{\infty}$$

$$f(s) = \frac{1}{s+a} \tag{3.6}$$

(*iv*) *Trigonometrical and hyperbolic functions*

Before we begin the derivation of the Laplace transforms of these functions, it should be noted that if two time functions $f_1(t)$ and $f_2(t)$ are added, the resulting sum has the Laplace transform

$$\mathcal{L}(f_1(t)+f_2(t)) = f_1(s)+f_2(s)$$

This can be proved by substitution into equation 3.1.

(*a*) $f(t) = \sin \omega t$

The Laplace transform of $\sin \omega t$ can be found by rewriting $\sin \omega t$ in terms of its exponential components and using equation 3.1.

$$f(s) = \int_{0-}^{\infty} \left(\frac{e^{j\omega t}-e^{-j\omega t}}{2j}\right) e^{-st}\, dt$$

$$f(s) = \frac{1}{2j}\left[\frac{e^{(-s+j\omega)t}}{-s+j\omega} - \frac{e^{(-s-j\omega)t}}{-s-j\omega}\right]_{0-}^{\infty}$$

$$f(s) = -\frac{1}{2j}\left(\frac{1}{-s+j\omega} - \frac{1}{-s-j\omega}\right)$$

$$f(s) = \frac{\omega}{s^2+\omega^2} \tag{3.7}$$

(*b*) $f(t) = \cos \omega t$

The Laplace transform of $\cos \omega t$ can be found in the same way as the Laplace transform of $\sin \omega t$.

$$\mathcal{L} \cos \omega t = \frac{s}{s^2+\omega^2} \tag{3.8}$$

(*c*) $f(t) = \sinh \omega t$

The Laplace transform of $\sinh \omega t$ can be found by rewriting $\sinh \omega t$ in terms of its exponential components and using equation 3.1.

$$f(s) = \int_{0-}^{\infty} \left(\frac{e^{\omega t}-e^{-\omega t}}{2}\right) e^{-st}\, dt$$

$$f(s) = \tfrac{1}{2}\left[\frac{e^{(-s+\omega)t}}{-s+\omega} - \frac{e^{(-s-\omega)t}}{-s-\omega}\right]_{0-}^{\infty}$$

$$f(s) = -\tfrac{1}{2}\left(\frac{1}{-s+\omega} - \frac{1}{-s-\omega}\right)$$

$$f(s) = \frac{\omega}{s^2-\omega^2} \tag{3.9}$$

(*d*) $f(t) = \cosh \omega t$

The Laplace transform of $\cosh \omega t$ can be found in the same way as the Laplace transform of $\sinh \omega t$.

$$\mathscr{L} \cosh \omega t = \frac{s}{s^2-\omega^2} \tag{3.10}$$

(*v*) *The unit step function and the delayed unit step function*

The unit step function is a step function with A equal to unity. It is given the symbol of $u(t)$ as a function of time.

A delayed unit step function is defined as follows

$$f(t) = u(t-T) = 0 \quad \text{when} \quad t < T$$

$$f(t) = u(t-T) = 1 \quad \text{when} \quad t > T$$

The definition gives a unit step function delayed by time $t = T$.

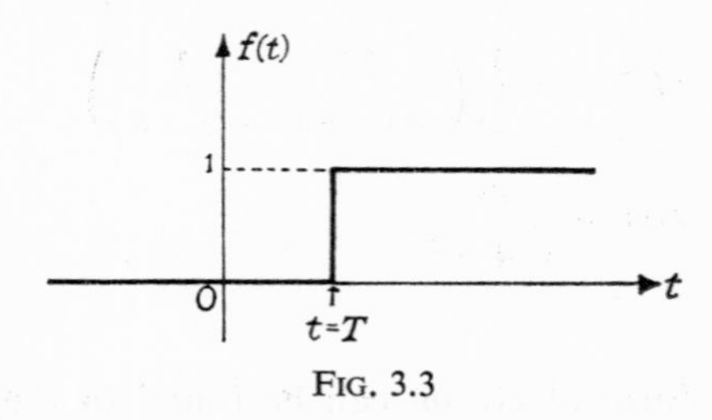

FIG. 3.3

Figure 3.3 shows the delayed unit step function under consideration. Using equation 3.1 and the definition of the delayed unit step function, from time $t = 0$, we can find its Laplace transform.

$$f(s) = \int_{0-}^{\infty} e^{-st} u(t-T)\, dt$$

$$f(s) = \int_{0-}^{T} e^{-st} u(t-T)\, dt + \int_{T}^{\infty} e^{-st} u(t-T)\, dt$$

By definition the delayed unit step function is zero between $t = 0$ and $t = T$, and is unity between $t = T$ and $t = \infty$, therefore

$$f(s) = \int_{T}^{\infty} e^{-st}\, dt$$

$$f(s) = \frac{e^{-sT}}{s} \tag{3.11}$$

(vi) The unit impulse function

The unit impulse function is an unusual but very useful mathematical function. It can be derived from a square pulse of unity area.

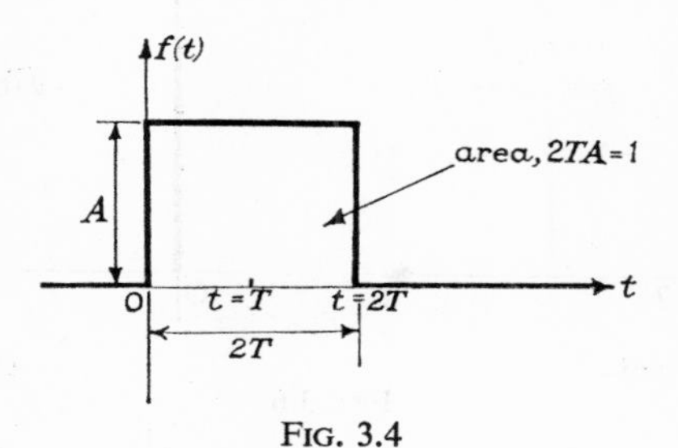

FIG. 3.4

Figure 3.4. The square pulse has a unity area if the product width × height, $2T \times A$, equals unity. The height of the pulse A is made much larger and the width much smaller, but the area $2TA$ is kept the same.

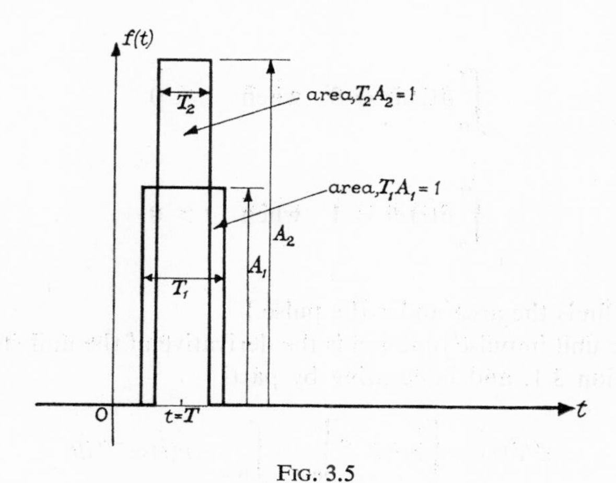

FIG. 3.5

Figure 3.5 shows examples of the change of the shape of the pulse. The pulse is centred at time $t = T$.

In the limit the pulse width tends to zero and the height tends to infinity. The pulse becomes an *impulse* of unit area. The function is called a *unit impulse*. It is given the notation

$$\delta(t-T)$$

which describes a unit impulse at time $t = T$.

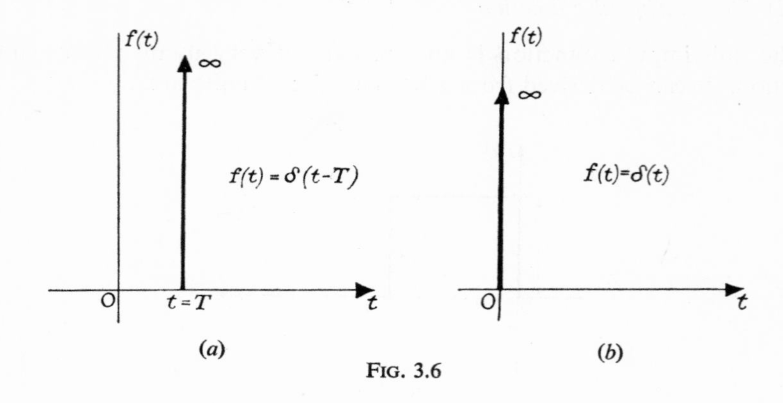

FIG. 3.6

Figure 3.6. (*a*) shows a unit impulse at time $t = T$, and (*b*) shows a unit impulse at time $t = 0$.

To find the Laplace transform of the unit impulse function we consider its integral,

$$\int_0^t \delta(t)\,dt = 0 \quad \text{when} \quad t < 0$$

$$\int_0^t \delta(t)\,dt = 1 \quad \text{when} \quad t > 0$$

(Integration finds the area under the pulse.)

Hence, the unit impulse function is the derivative of the unit step function. Using equation 3.1, and integrating by parts

$$\mathscr{L}\delta(t) = \left[u(t)e^{-st}\right]_{0-}^{\infty} - \int_{0-}^{\infty} -u(t)se^{-st}\,dt$$

$$\mathscr{L}\delta(t) = 0 - \left[e^{-st}\right]_{0-}^{\infty}$$

$$\mathscr{L}\delta(t) = 1 \tag{3.12}$$

The unit impulse function is often called the *Dirac* or *delta function*.

More Laplace transforms of functions may be derived using equation 3.1. The following table shows the more useful ones.

Table of Laplace Transforms of Functions

	$f(s) = \mathscr{L}f(t)$	$f(t)$
1	1	$\delta(t)$
2	$\frac{1}{s}$	$u(t)$
3	$\frac{1}{s^2}$	t
4	$\frac{n!}{s^{n+1}}$	t^n
5	$\frac{1}{s+a}$	e^{-at}
6	$\frac{1}{(s+a)^2}$	te^{-at}
7	$\frac{n!}{(s+a)^{n+1}}$	$t^n e^{-at}$
8	$\frac{1}{s(s+a)^2}$	$\frac{1}{a^2}(1-(1+at)\,e^{-at})$
9	$\frac{1}{(s+a)(s+b)}$	$\frac{e^{-at}-e^{-bt}}{b-a}$
10	$\frac{\omega}{s^2+\omega^2}$	$\sin \omega t$
11	$\frac{s}{s^2+\omega^2}$	$\cos \omega t$
12	$\frac{\omega}{s^2-\omega^2}$	$\sinh \omega t$
13	$\frac{s}{s^2-\omega^2}$	$\cosh \omega t$
14	$\frac{s+a}{s^2+\omega^2}$	$\frac{\sqrt{(a^2+\omega^2)}}{\omega}\sin(\omega t+\tan^{-1}\omega/a)$
15	$\frac{\omega}{(s+a)^2+\omega^2}$	$e^{-at}\sin \omega t$
16	$\frac{s+a}{(s+a)+\omega^2}$	$e^{-at}\cos \omega t$
17	$\frac{1}{s^2+2\zeta\omega_n s+\omega_n^2}$	$\frac{1}{\omega_n\sqrt{(1-\zeta^2)}}\,e^{-\zeta\omega_n t}\sin \omega_n\sqrt{(1-\zeta^2)}\,t$
18	$\frac{1}{s(s^2+2\zeta\omega_n s+\omega_n^2)}$	$\frac{1}{\omega_n^2}\left\{1-\frac{1}{\sqrt{(1-\zeta^2)}}\,e^{-\zeta\omega_n t}\sin[\omega_n\sqrt{(1-\zeta^2)}\,t+\phi]\right\}$ where $\phi = \cos^{-1}\zeta$

Operations

In the following consideration of the Laplace transforms of operations the variable x is a function of time t.

(*i*) *Real Differentiation*

(*a*) Let $f(t) = \dfrac{dx}{dt}$

The Laplace transform of dx/dt can be found by using equation 3.1.

$$f(s) = \int_{0-}^{\infty} \frac{dx}{dt}\, e^{-st}\, dt$$

integrating by parts

$$f(s) = \Big[x\, e^{-st}\Big]_{0-}^{\infty} - \int_{0-}^{\infty} -sx\, e^{-st}\, dt$$

$$f(s) = -x(0-)+s\bar{x} \qquad (3.13)$$

In equation 3.13 $x(0-)$ is the value of xe^{-st} at $t = 0-$ and the value of xe^{-st} at $x = \infty$ is zero. $x(0-)$ is the initial value of x. ($\bar{x}$ indicates that the original x has now become a function of the complex variable s.)

(*b*) Let $f(t) = \dfrac{d^2x}{dt^2}$

The Laplace transform of d^2x/dt^2 can be found by using equation 3.1 and integrating by parts.

$$f(s) = \int_{0-}^{\infty} \frac{d^2x}{dt^2}\, e^{-st}\, dt$$

$$f(s) = s^2\bar{x}-sx(0-)-\left[\frac{dx}{dt}\right]_{t=0-} \qquad (3.14)$$

(*c*) Let $f(t) = \dfrac{d^nx}{dt^n}$

The Laplace transform of the nth derivative of x follows from (*a*) and (*b*).

$$f(s) = s^n\bar{x}-s^{n-1}x(0-)-s^{n-2}\left[\frac{dx}{dt}\right]_{t=0-} \cdots \qquad (3.15)$$

In equation 3.15 the terms on the right-hand side after $s^n\bar{x}$ are all initial values. Usually when considering control systems all the initial values are zero and equation 3.15 becomes

$$f(s) = s^n\bar{x} \qquad (3.16)$$

(*ii*) *Real Integration*

Let $f(t) = \int_{0-}^{t} x\,dt + x_1$ (x_1 is the constant of integration). The Laplace transform of $f(t)$ can be found by integrating by parts and using equation 3.1.

$$f(s) = \int_{0-}^{\infty} e^{-st}\left(\int_{0-}^{t} x\,dt\right) dt + \frac{x_1}{s}$$

$$f(s) = \left[-\frac{1}{s}e^{-st}\int_{0-}^{t} x\,dt\right]_{0-}^{\infty} - \frac{1}{s}\int_{0-}^{\infty} -e^{-st}x\,dt + \frac{x_1}{s}$$

$$f(s) = \frac{\bar{x}}{s} + \frac{x_1}{s} \tag{3.17}$$

x_1 is the initial value of the integral, i.e. the value of the integral at the origin approached from the $t-$ side. For zero initial conditions

$$f(s) = \frac{\bar{x}}{s}$$

Note

When a function of the complex variable s is transformed into a function of time t *inverse transformation* is said to take place. The inverse transformation is usually indicated by the symbol $\mathscr{L}^{-1}$.

3C The use of the Laplace transform in solving linear differential equations

The following worked examples illustrate the use of Laplace transforms in solving linear differential equations.

Example (*i*)

Find the value of x, when at all times t is greater than zero, given that

$$\tfrac{1}{3}\frac{d^2x}{dt^2} + \tfrac{16}{3}x = \delta(t) \tag{3.18}$$

The initial value of x is 1 and the initial value of dx/dt is zero.

The Laplace transform of both sides of equation 3.18 is

$$\underbrace{\tfrac{1}{3}\left(s^2\bar{x} - sx(0-) - \left[\frac{dx}{dt}\right]_{t=0-}\right)}_{\mathscr{L}\frac{1}{3}\frac{d^2x}{dt^2}} + \underbrace{\tfrac{16}{3}\bar{x}}_{\mathscr{L}\frac{16}{3}x} = \underbrace{1}_{\mathscr{L}\delta(t)} \tag{3.19}$$

Substituting the initial values into equation 3.19 we have

$$s^2\bar{x}-s+16\bar{x} = 3$$

$$(s^2+16)\bar{x} = s+3$$

$$x = \mathscr{L}^{-1}\frac{s+3}{s^2+16} \tag{3.20}$$

The inverse transform of equation 3.20 can be found by referring to number 14 in the table of Laplace transforms, therefore

$$x = \tfrac{5}{4}\sin(4t+0{\cdot}927)$$

Example (ii)

A 1 μF capacitor and an inductor of resistance 100 Ω and inductance 2·5 mH are connected in series. A d.c. voltage of 100 V is applied to the series combination at time $t = 0$. Find the current flowing in the circuit at any time t after $t = 0-$. Initial conditions are zero.

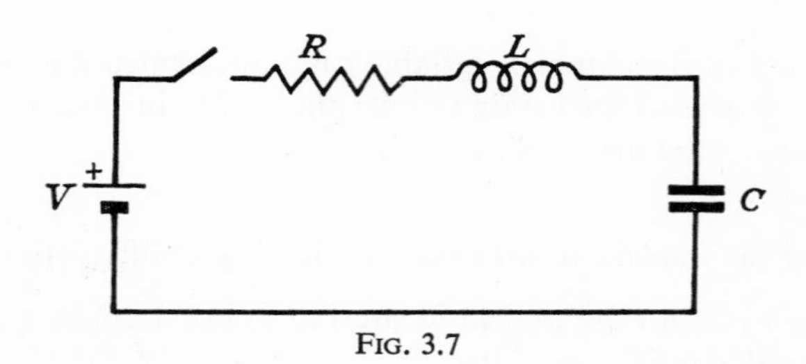

FIG. 3.7

Figure 3.7 shows the series circuit under consideration. Let the current and voltage as functions of time be i and v respectively. Applying Kirchhoff's second law we have

$$v = Ri+L\frac{di}{dt}+\frac{1}{C}\int_0^t i\,dt \tag{3.21}$$

v is a step function voltage of 100 V. The Laplace transform of both sides of equation 3.21 is

$$\frac{100}{s} = \frac{1}{sC}\bar{\imath}+R\bar{\imath}+Ls\bar{\imath}$$

$$\frac{100}{s} = \left(\frac{1}{s\,10^{-6}}+10^2+2{\cdot}5\times10^{-3}s\right)\bar{\imath}$$

therefore
$$\bar{i} = \left(\frac{4\times10^4}{s^2+4\times10^4 s+4\times10^8}\right)$$

$$i = \mathscr{L}^{-1}\frac{4\times10^4}{(s+2\times10^4)^2} \tag{3.22}$$

The inverse transform of equation 3.22 can be found by referring to number 6 in the table of Laplace transforms, therefore

$$i = 4\times10^4 t\,\mathrm{e}^{-2\times10^4 t} \qquad \text{(A)}$$

Example (iii)

The position of a flywheel is controlled by a remote position control system. The system moment of inertia is 100 kg-m^2 and the motor torque applied to the flywheel is 2500 N-m per radian of misalignment between the output and input position. A flywheel velocity of 1 radian per second produces a feed-back torque on the flywheel of 600 N-m. The input position is suddenly turned $\pi/2$ radians at time $t = 0$. Find an expression for the position of the flywheel at any time t after $t = 0$.

The basic equation of the system is

$$\frac{J}{K}\frac{d^2\theta_o}{dt^2}+\frac{F}{K}\frac{d\theta_o}{dt}+\theta_o = \theta_i \tag{3.23}$$

where θ_o and θ_i are output and input positions respectively, J is the system inertia, F is the velocity feed-back torque and K is the torque per radian of misalignment.

All initial conditions are zero. Taking the Laplace transform of both sides of equation 3.23 and substituting the given values we obtain

$$\tfrac{100}{2500}\,s^2\bar{\theta}_o+\tfrac{600}{2500}\,s\bar{\theta}_o+\bar{\theta}_o = \frac{\pi/2}{s} \tag{3.24}$$

Equation 3.24 can be rearranged using partial fractions (chapter 2, section 2A), giving

$$\bar{\theta}_o = \frac{\pi}{2}\left(\frac{1}{s}-\frac{s+6}{(s+3)^2+16}\right) \tag{3.25}$$

The inverse transform of equation 3.25 can be found by referring to number 18 in the table of Laplace transforms, therefore

$$\theta_o = \frac{\pi}{2}\,(1-\tfrac{5}{4}\,\mathrm{e}^{-3t}\sin(4t+0{\cdot}927))\ \text{radians}$$

3D Comments

Students who have read this chapter and chapter 2 should notice the similarity between the p-operator and s, the Laplace variable. s and p may be assumed identical in considering control systems, initial conditions usually being zero.

The 'mathematics' in chapter 2 was convenient. In this chapter, however, we have been more rigorous. It is presented for engineers who wish to use their mathematics. For those who are interested in a thorough approach to Laplace transforms the following references may be useful.

1. KUO, F. F.: *Network Analysis and Synthesis* (Wiley, 1962).
2. CARSLAW, H. S. and JAEGER, J. C.: *Operational Methods in Applied Mathematics* (Oxford University Press, 1949; now available in Dover reprint from Constable).

3E Examples

1. Using the definition of the Laplace transform find the transforms of the following functions of time t:

(*i*) $f(t) = 3t^3$
(*ii*) $f(t) = e^{-3t} \sin 5t$
(*iii*) $f(t) = 1{\cdot}25 \sin(4t+0{\cdot}927)$
(*iv*) $f(t) = \dfrac{e^{-3t}}{2}(e^{-2t}-e^{-4t})$
(*v*) $f(t) = 10 \cosh 2t$

$$\left(\frac{18}{s^4};\ \frac{5}{s^2+6s+34};\ \frac{s+3}{s^2+16};\ \frac{1}{s^2+12s+35};\ \frac{10s}{s^2-4}\right)$$

2. Using the definition of the Laplace transform find the transforms of the following operations. (The variable is a function of time t.)

(*i*) $f(t) = \dfrac{dx^3}{dt^3}$ where $x(0-) = 1$ and $\left[\dfrac{dx}{dt}\right]_{t=0-} = \left[\dfrac{d^2x}{dt^2}\right]_{t=0} = 0$

(*ii*) $f(t) = 3\dfrac{dx}{dt} - \dfrac{d^2x}{dt^2}$ where $x(0-) = 2$ and $\left[\dfrac{dx}{dt}\right]_{t=0-} = 1$

$$[s^3\bar{x}-s^2;\ (3s-s^2)\bar{x}+2s-5]$$

3. An electric circuit consists of a 50 μF capacitor connected in series with a 2 kΩ resistor. Initially the capacitor is charged to 100 μC. If a d.c. voltage of 10 V is suddenly applied to the circuit, find an expression for the variation of voltage across the capacitor.

$$[10(1-0{\cdot}8\, e^{-10t})]$$

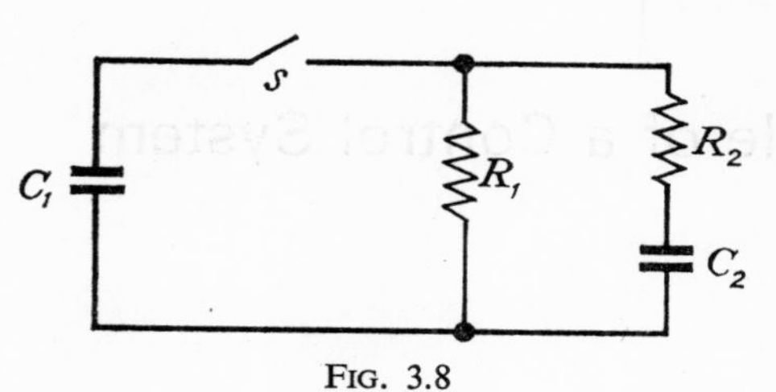

FIG. 3.8

4. In the figure the capacitor C_1 is charged to 100 V and then discharged by the sudden closing of the switch S. Derive an expression for the subsequent variation with time of the voltage across C_2 (which is initially uncharged), if $C_1 = 0{\cdot}02\ \mu F$, $C_2 = 0{\cdot}0025\ \mu F$, $R_1 = 10\ k\Omega$ and $R_2 = 2 k\Omega$.

(Part 3. I.E.E. 1957)

$[900(e^{-0{\cdot}05\times 10^5 t} - e^{-2{\cdot}25\times 10^5 t})]$

CHAPTER 4

An Example of a Control System

*4.1 The remote position control system

It should be remembered that the requirement of a remote position control system is that, ideally, the output shaft's position should be the same as the input shaft's position. Consider the following remote position control system.

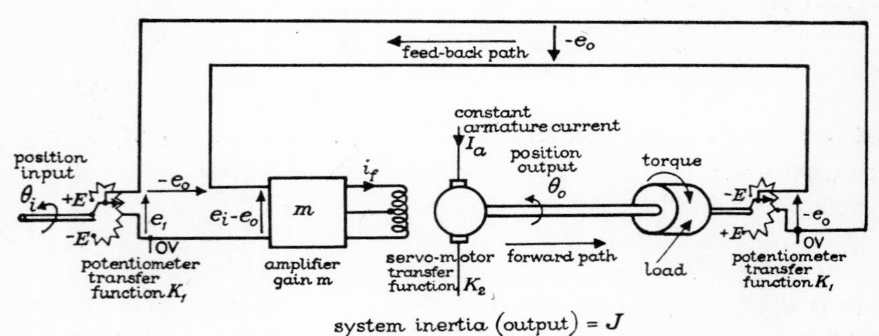

FIG. 4.1

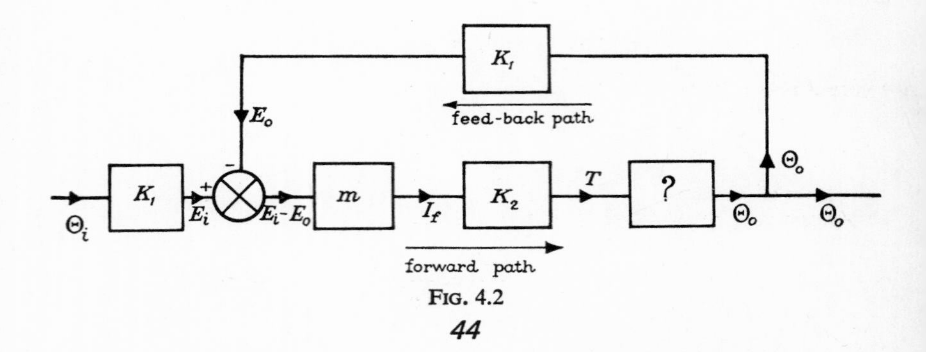

FIG. 4.2

Figure 4.1 shows a practical remote position control system. The figure becomes much simpler if the individual components are replaced by their block diagrams and specified by their transfer functions.

Figure 4.2 shows the block diagram. The various component transfer functions are as follows:

$$K_1 = \frac{e_i}{\theta_i} \tag{4.1}$$

where e_i volts is the potential across the potentiometer applied to the amplifier. θ_i radians is the input shaft position, K_1 is constant, the potentiometer being linear.

$$m = \frac{i_f}{e_i - e_o} \tag{4.2}$$

where i_f amperes is the output current of the amplifier applied to the field of the servo-motor. $-e_o$ volts is the feed-back signal. $e_i - e_o$ is in fact the error. (It should be recalled that if $e_i = e_o$ no signal is applied to the system as it should have the correct output.) m is constant, the amplifier being used over a linear portion of its characteristic.

$$K_2 = \frac{T}{i_f} \tag{4.3}$$

where T is the motor torque and hence the load output torque. K_2 is a constant, the output torque being proportional to field current.

$$? = \frac{\theta_o}{T} \tag{4.4}$$

This is a hypothetical component which transfers output torque to output position θ_o. In fact all components attached to the output contribute to the load on the output and hence affect the inertia of the load. Using the relationship

$$\text{output torque} = \text{accleration} \times J$$

where J is the system inertia we have

$$T = \frac{d^2\theta_o}{dt^2} J$$

Expressed in p-operator notation we have

$$\frac{\Theta_o}{T} = \frac{1}{Jp^2}$$

$$K_1 = \frac{E_o}{\Theta_o} \tag{4.5}$$

The output potentiometer is calibrated identically with the input potentiometer.

Simplifying the block diagram as follows:

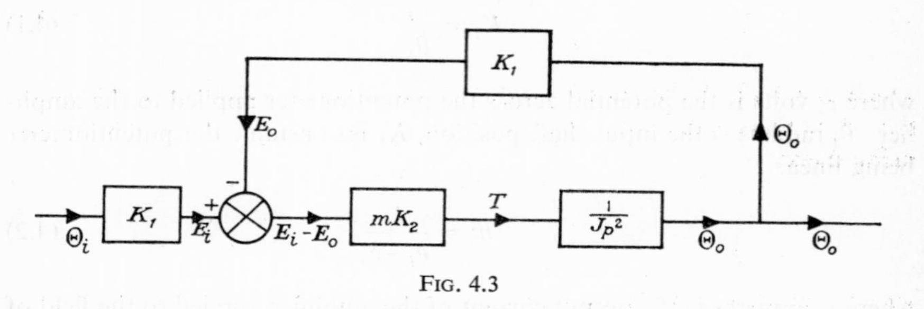

FIG. 4.3

Figure 4.3 shows the simplified block diagram now consisting of four blocks, they can all be expressed in system parameters.

$$\frac{mK_2}{Jp^2} = \frac{\Theta_o}{E_i - E_o} \tag{4.6}$$

$$K_1 = \frac{E_i}{\Theta_i} = \frac{E_o}{\Theta_o}$$

Therefore $E_i = K_1\Theta_i$ and $E_o = K_1\Theta_o$. These expressions are substituted into equation 4.6 to obtain an expression, which consists of only input and output position as functions of the p-operator,

$$\frac{mK_2}{Jp^2} = \frac{\Theta_o}{K_1(\Theta_i - \Theta_o)} \tag{4.7}$$

By rearranging equation 4.7 we may obtain the complete transfer function (Θ_o/Θ_i) of the system

$$J \times p^2\Theta_o = mK_1K_2 \times (\Theta_i - \Theta_o)$$

$$\begin{pmatrix}\text{system}\\ \text{inertia}\end{pmatrix} \times \begin{pmatrix}\text{load angular}\\ \text{acceleration}\end{pmatrix} = \begin{pmatrix}\text{output torque}\\ \text{per radian of}\\ \text{error}\end{pmatrix} \times \begin{pmatrix}\text{difference between}\\ \text{output and input}\\ \text{position error}\end{pmatrix} = \text{output torque}$$

Further rearrangement of equation 4.7 gives

$$\Theta_o = \Theta_i\left(\frac{mK_1K_2/J}{p^2 + mK_1K_2/J}\right) \tag{4.8}$$

which is a second-order linear differential equation. Our ideal system should give us

$$\Theta_o = \Theta_i$$

but we can see from equation 4.8 that this is not so with this particular system. However the solution of this equation may indicate a useful close approximation. Now we can make any input we wish and, using the operator methods already discussed, solve for θ_o.

Make θ_i a step function input specified by

$$\theta_i = 0 \text{ radians before time } t = 0$$
$$\theta_i = A \text{ radians after time } t = 0$$

At time $t = 0$ the output position $\theta_o = 0$ (initially the system is correct).

Substituting the operator form, A/p, of the input into equation 4.8 gives

$$\Theta_o = \frac{A}{p}\left[\frac{(mK_1K_2)/J}{p^2+mK_1K_2/J}\right]$$

To solve for the output θ_o this equation is expressed in partial fractions,

$$\Theta_o = A\left[\frac{B_1}{p}+\frac{B_2p+B_3}{p^2+(mK_1K_2)/J}\right]$$

where B_1, B_2 and B_3 are the partial fraction constants and are found to be 1, -1, and 0 respectively. Substituting these values into the previous equation, we obtain

$$\Theta_o = A\left[\frac{1}{p}-\frac{p}{p^2+(mK_1K_2)/J}\right] \tag{4.9}$$

We can change equation 4.9 into an equation as a function of time by referring to 'equations' 2.7 and 2.12 in the table of operator notation, page 23, giving

$$\theta_o = A\,\{1-\cos\sqrt{[(mK_1K_2)/J]}\,.\,t\} \tag{4.10}$$

Let $mK_1K_2 = K_3$, thus equation 4.10 becomes

$$\theta_o = A\,(1-\cos\sqrt{(K_3/J)}\,.\,t) \tag{4.11}$$

Equation 4.11 describes the behaviour of the output θ_o for an input step function. Obviously this is not the required behaviour of a remote position control system. To examine the output variation compared with the input we shall draw a graph showing output θ_o and input θ_i to the base of time t.

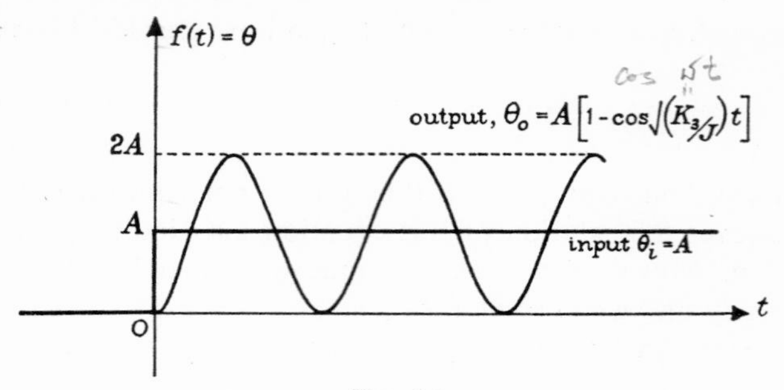

Fig. 4.4

Figure 4.4 shows that although the output is required to be a steady value of A radians it is oscillating with a peak amplitude of $2A$ radians. This system is, of course, useless, but now that we know something of its behaviour it can be corrected.

To correct the system the oscillations must be damped down somehow. The frequency of oscillation of the output is $\sqrt{(K_3/J)}$ radians per second. It is called the undamped natural frequency of the system. (The system is obviously undamped because equation 4.11 indicates that after time $t = 0$ the system output will continue oscillating at the same amplitude.) The undamped natural frequency is given the symbol ω_n when expressed in radians per second, or f_n when expressed in hertz. One means of damping the oscillations is by attaching to the output shaft a vane that is free to rotate in oil.

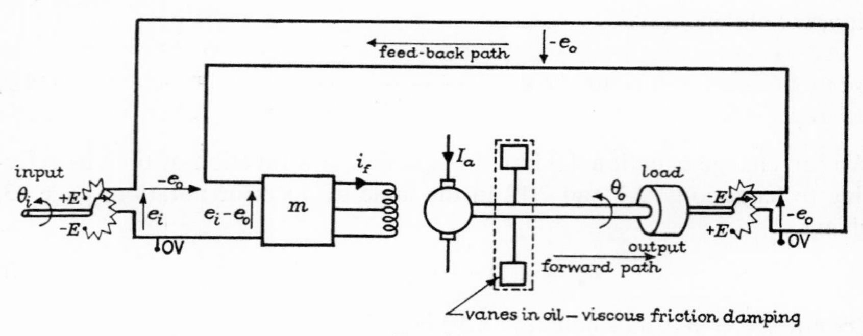

Fig. 4.5

Figure 4.5 shows a remote position control system with a damping device attached to the output shaft. This is called *viscous friction damping* and provides a friction force which produces a torque opposing the motion of rotation of the shaft. The torque varies as the output shaft speed. This effect may be demonstrated by running one's hand through a bath of water—as the speed of the hand is increased, the viscous force exerted by the water on the hand also increases.

Now let us examine whether the introduction of this device improves the system or not. The equation of the system before the introduction of viscous friction was

$$Jp^2\Theta_o = \underbrace{K_3(\Theta_i - \Theta_o)}_{\text{torque}} \tag{4.12}$$

An opposing torque T_F has now been introduced and equation 4.12 is modified to

$$Jp^2\Theta_o = K_3(\Theta_i - \Theta_o) - \underset{\text{modifying torque}}{T_F} \tag{4.13}$$

The torque T_F is produced by a frictional force proportional to the output shaft speed and therefore

$$T_F = F\frac{d\theta_o}{dt} \tag{4.14}$$

where F is a constant. In operator form equation 4.14 is

$$T_F = Fp\Theta_o \tag{4.15}$$

Substituting equation 4.15 into equation 4.13 and rearranging we have

$$\underset{\text{inertia}\times\text{acceleration}}{Jp^2\Theta_o} = \underset{\text{output torque}}{K_3(\Theta_i - \Theta_o)} - \underset{\text{opposing torque due to viscous friction}}{Fp\Theta_o} \tag{4.16}$$

Again, we must examine whether the output position Θ_o either equals the input position Θ_i at all times, as ideally required, or whether it usefully approaches this condition for any values of F and K_3 (these are variable parameters in a system). First, for ease of analysis, rearrange equation 4.16 as follows

$$\frac{J}{K_3}p^2\Theta_o + \frac{F}{K_3}p\Theta_o + \Theta_o = \Theta_i \tag{4.17}$$

Now put $\sqrt{(K_3/J)} = \omega_n$.

Substituting into equation 4.17 we have

$$\frac{1}{\omega_n^2}p^2\Theta_o + \frac{F}{K_3}p\Theta_o + \Theta_o = \Theta_i$$

therefore

$$\Theta_o = \Theta_i\left(\frac{1}{p^2/\omega_n^2 + p\,F/K_3 + 1}\right) \tag{4.18}$$

To solve equation 4.18 we will use the same step input ($\Theta_i = A/p$) as before and we will arrange for all initial conditions to be zero. Therefore

$$\Theta_o = \frac{A}{p}\left(\frac{1}{p^2/\omega_n^2+p\,F/K_3+1}\right) \tag{4.19}$$

Equation 4.19 has as a characteristic part the quadratic $p^2/\omega_n^2+p\,F/K_3+1$. (This is called the characteristic equation.) This equation can be factorized and will have roots of the values

$$-\frac{F\omega_n^2}{2K_3}\pm\sqrt{\left(\frac{F^2\omega_n^4}{4K_3}-\omega_n^2\right)} \tag{4.20}$$

These roots will be either real, complex, or equal depending on the values of F, K_3 and ω_n. So it can be seen that there are three possible complete solutions of equation 4.17 and that the modified remote position control system has three types of behaviour. For simplicity let

$$\frac{F\omega_n}{2K_3} = \zeta$$

The expression for the roots of equation 4.20 becomes

$$-\zeta\omega_n \pm \omega_n\sqrt{(\zeta^2-1)}$$

(*i*) For real roots when $\zeta^2 > 1$ (i.e. $\zeta > 1$) the solution for the output shaft position is

$$\Theta_o = A\left\{1-\frac{e^{-\zeta\omega_n t}}{\sqrt{(\zeta^2-1)}}\sinh\left[\omega_n\sqrt{(\zeta^2-1)}\,t+\phi\right]\right\} \text{ where } \phi = \cosh^{-1}\zeta \tag{4.21}$$

(*ii*) For imaginary roots when $\zeta < 1$ the solution for the output shaft position is

$$\theta_o = A\left\{1-\frac{e^{-\zeta\omega_n t}}{\sqrt{(1-\zeta^2)}}\sin\left[\omega_n\sqrt{(1-\zeta^2)}\,t+\phi\right]\right\} \text{ where } \phi = \cos^{-1}\zeta \tag{4.22}$$

(*iii*) For equal roots when $\zeta = 1$ the solution for the output shaft position is

$$\theta_o = A\left[1-(1+\omega_n t)e^{-\omega_n t}\right] \tag{4.23}$$

The mathematics involved in obtaining the three solutions is tedious and would certainly distract from the physical interpretation of the system, but an example of the solution of this form of equation is shown in chapter 2, section 2.3. The best way to interpret the solutions is to draw graphs of them and to compare them with the input.

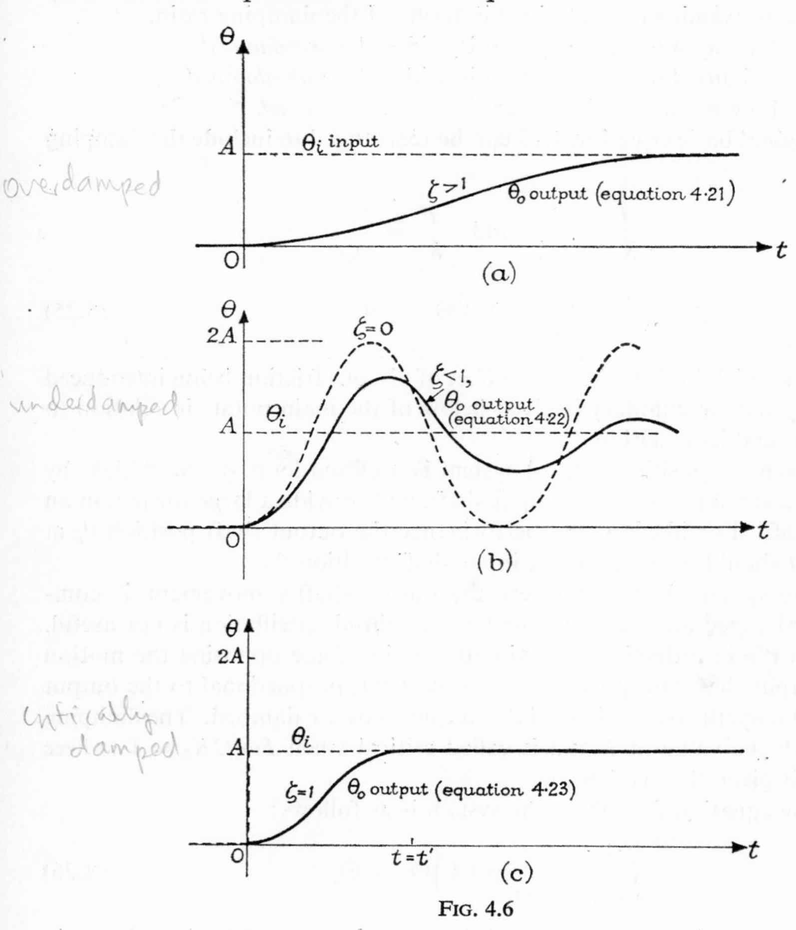

FIG. 4.6

Figure 4.6 (*a*), (*b*) and (*c*) shows the graphs of the solutions (for particular and unspecified values of F, K_3 and ω_n). It can be seen that after a time t' output (*c*) gives a shaft position exactly the same as the position of the input shaft. Thus, when $\zeta = 1$ a reasonably accurate system is obtained, and as

$$\zeta = \frac{F\omega_n}{2K_3}$$

$$\frac{F\omega_n}{2K_3} = 1 \tag{4.24}$$

Equation 4.24 is said to indicate critical damping. ζ is called the *damping ratio.*

Now let us examine our solutions in terms of the damping ratio.

(*a*) $\zeta > 1$ figure 4.6(*a*), the system is said to be *overdamped.*

(*b*) $\zeta < 1$ figure 4.6(*b*), the system is said to be *underdamped.*

(*c*) $\zeta = 1$ figure 4.6(*c*), the system is *critically damped.*

The original basic equation 4.18 can be rearranged to include the damping ratio.

As
$$\frac{J}{K_3} = \frac{1}{\omega_n^2} \quad \text{and} \quad \frac{F}{K_3} = \frac{2\zeta}{\omega_n}$$

we have
$$\frac{p^2\Theta_o}{\omega_n^2} + \frac{2\zeta}{\omega_n} p\Theta_o + \Theta_o = \Theta_i \qquad (4.25)$$

Taking a quick look again at the effect of viscous friction being introduced into the system, a summary is given below of the main points in relation to the system and its equation.

(*i*) A remote position control system is a closed-loop system which, by means of a small torque on an input shaft, will provide a large torque on an output shaft. For ideal system performance the output shaft position θ_o at any time t should be equal to the input shaft position θ_i.

(*ii*) The system described where the output shaft's movement is completely undamped and hence of constant-amplitude oscillation is not useful.

(*iii*) On the introduction of a viscous friction force opposing the motion of the output shaft and providing a torque $Fp\Theta_o$ proportional to the output shaft's velocity, the oscillations of the output shaft are damped. The damping varies with the value of F and is called critical when $F\omega_n/2K_3 = 1$ (where $F\omega_n/2K_3$ is given the symbol ζ).

(*iv*) The equation describing the system is as follows:

$$\left(\frac{1}{\omega_n^2} p^2 + \frac{2\zeta}{\omega_n} p + 1\right)\Theta_o = \Theta_i \qquad (4.26)$$

(*v*) Interpretation of the system performance in terms of the damping ratio for an input step function is as follows:

$\zeta = 0$, the system is undamped giving an oscillatory output. The system is unstable.

$\zeta \ll 1$, the system is very underdamped and the output oscillates for some time before settling down to the same magnitude as the input.

$\zeta < 1$, the system is underdamped, but the output oscillations decrease fast enough for the system output to become useful.

$\zeta = 1$, the system is critically damped, the output rises to the same magnitude as the input with no overshoot.

$\zeta > 1$, the system is overdamped and the output takes a comparatively long time before it reaches the same magnitude as the input.

(*vi*) In practical remote position control systems it is usual to design the system such that ζ is some value between 0·4 and 0·9.

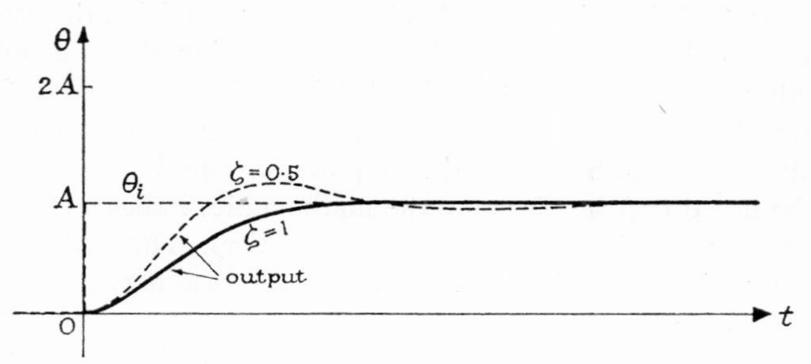

FIG. 4.7

Figure 4.7 shows that in the case of $\zeta = 0{\cdot}5$ a slight overshoot will occur, but the advantage of this damping ratio rather than critical damping is that the system responds faster.

(*vii*) Friction in the various mechanical parts of the control system has so far been ignored. Assuming the output shaft to be at rest there will, initially, be a static torque to overcome. When the shaft is moving there will be a constant rubbing friction causing a constant small friction torque, which can for present theoretical and practical purposes be ignored. The initial 'static' friction is generally unpredictable and introduces an unreliability in obtaining a predictable output shaft position when the output shaft comes to rest. This is a non-linear effect and is dealt with in more detail in chapter 11. However, with carefully machined bearings the inaccuracy is removed as much as possible.

The use of viscous friction damping by means of a vane immersed in oil is not the most satisfactory technique for a remote position control system. The effect of viscous friction can, however, be produced in another way. Let us look at the basic equation of the system damped by viscous friction,

$$\frac{p^2\Theta_o}{\omega_n^2}+\frac{2\zeta}{\omega_n}\,p\Theta_o+\Theta_o = \Theta_i$$

Rearranging and substituting J/K_3 for $1/\omega_n^2$ and mK_1K_2 for K_3

$$Jp^2\Theta_o = mK_1K_2\Theta_i-mK_1K_2\Theta_o-mK_1K_2\,\frac{2\zeta}{\omega_n}\,p\Theta_o$$

$$Jp^2\Theta_o = mK_2\left[(K_1\Theta_i-K_1\Theta_o)-K_1\,\frac{2\zeta}{\omega_n}\,p\Theta_o\right]$$

but $K_1\Theta_i = E_i$ and $K_1\Theta_o = E_o$

therefore
$$Jp^2\Theta_o = mK_2\left[(E_i - E_o) - K_1\frac{2\zeta}{\omega_n}p\Theta_o\right] \tag{4.27}$$

In equation 4.27 there is a term $K_1(2\zeta/\omega_n)p\Theta_o$ modifying the error potential difference and opposing it. This term, dimensionally, is a potential (if in doubt about the dimensions, a simple dimensional analysis of equation 4.27, bearing in mind that $p = d/dt =$ [1/time] will show this). The error potential difference can be modified by a potential equal to $-K_1(2\zeta/\omega_n)p\Theta_o$. It should be noted that just because the mathematics indicates that a system like this can 'exist' does not prove that it is practical. The engineer may find a system impossible to design, but this is not the case here as the following will demonstrate.

A tacho-generator is a device which, when attached to a rotating shaft, gives an electrical output proportional to the speed of the shaft (e.g. a d.c. generator with a constant field).

$$\text{tacho-generator output} = \text{constant} \times \underbrace{p\Theta}_{\text{shaft velocity}}$$

Thus if a tacho-generator is attached to the output shaft of the remote position control system its output potential will be a constant multiplied by $p\Theta_o$. If the constant is arranged to be $K_1(2\zeta/\omega_n)$ and the output is connected in the system such that it opposes the error potential difference $E_i - E_o$, viscous friction damping will now have been simulated. This is shown as follows:

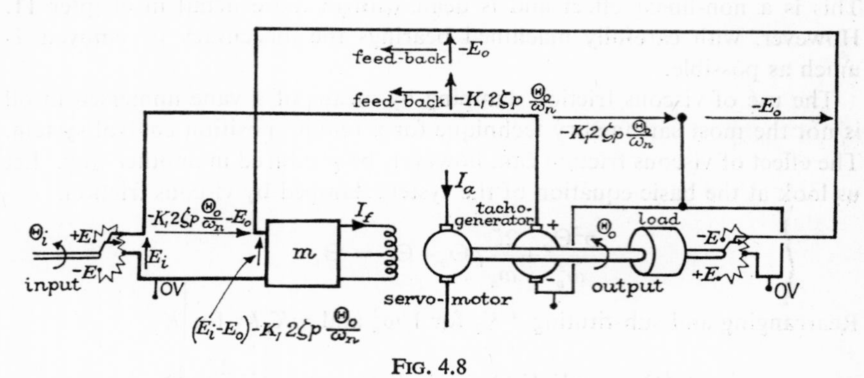

FIG. 4.8

Figure 4.8 shows this arrangement and the system is now a remote position control system with output velocity feed-back (sometimes called output derivative feed-back).

*4.2 Stability and linearity

Stability

It seems obvious to say that a control system should be stable, but it cannot be over-emphasized that when a control system is designed the engineer must make certain that the output is absolutely stable (response to a step function input is a suitably bounded output) under all required input conditions. The problem of stabilization is not simple, as will no doubt become evident from the amount of text devoted to it in this book.

The remote position control system discussed without damping seems superficially to have all the components required to give an accurate output, but closer examination shows it to be unstable and hence useless. This is but a simple system, so it may be realized that the problems involved in analysing more complicated systems are not straight-forward.

Linearity

In previous sections all components of the simple remote position control system were stated to be linear over the range of operation (i.e. transfer functions of individual components do not vary).† Although, for analysis and prediction of behaviour, linear engineering components are usually desirable, linearity is not natural.

Processes are often non-linear, having variable 'transfer functions' depending either on input variation or surrounding conditions, such as the temperature of the atmosphere.

Most of the control system theory in this book is based on linear theory and the examination of non-linear theory, although becoming increasingly important, is complicated and still in its infancy.

Chapter 11 discusses non-linearities.

*4.3 Comments

(*i*) The output of a control system should, ideally, be the same as the calibrated input, but short periods of difference can, in some systems, be tolerated.

(*ii*) The remote position control system is used as an example throughout this text. The following list gives some of the many fields in which control systems are used:

General automation in industry
Control of nuclear reactors

† This is not necessarily true, as the components are slightly non-linear, but linearity can be assumed because the non-linearities are so small.

Regulators { Speed control
Pressure control
Temperature control, etc.

{ Aircraft, missiles and 'space engineering'
Autopilots

Control of communication antennas (radar, etc.)

(*iii*) The block diagram and the transfer function give respectively a pictorial and a mathematical description of a control system and are useful 'tools' in the analysis and synthesis of systems in general.

In this chapter the behaviour and mathematics of the remote position control system example are often obtained by observation and not by a detailed use of block diagrams and transfer functions. The idea of introducing block diagrams and transfer functions at this stage is to familiarize students with these techniques for describing systems.

(*iv*) The step function input to the remote position control system is not a practical input; it is chosen as a useful function because of its mathematical description. It does, of course, give accurate and useful information.

4.4 Examples

1. Comment on the significance of the error signal in a remote position control system. Briefly comment on how a remote position control system is error-actuated.

2. Discuss the need for damping in a remote position control system and give an example of a critically damped system.

3. Show that an undamped remote position control system is unstable.

4. An error-actuated electrical servo-mechanism is employed to control the angular position of a rotatable mass, subject to viscous damping, in response to the rotation of a control handle. Give a circuit diagram of a suitable scheme and set up the equation of motion of the system.

In a particular case the moment of inertia of the rotating mass under control is 2000 kg-m^2 and the system is critically damped. The motor produces a torque of 1800 newton-metres per minute misalignment. Calculate the steady-state angular error if the control handle is continuously rotated at a speed of 2 r.p.m.

(Part 3. I.E.E. 1957)

(0·0075 radian)

5. A servo system for the positional control of a rotatable mass is stabilized by viscous damping, which is less than that required for critical damping. Derive an expression for the output of the system if the input member is suddenly moved to a new position, the system being initially at rest.

Calculate the amount of the first overshoot if the undamped natural frequency is 5 c/s and the frictional forces are one-third of the forces required for critical damping.

(Part 3. I.E.E. 1957)

(31·8%)

6. The angular position of a rotatable mass is controlled from a hand-wheel by means of a closed-loop electrical servo-mechanism, which is critically damped by viscous friction. Set up the differential equation of the system. Derive an expression for the angular position of the mass at any time if, with the system initially at rest, the control wheel is suddenly turned through an angle θ_1.

Given that the moment of inertia of the moving parts is 500 kg-m^2 and the motor torque is 2000 N-m per radian of misalignment and that $\theta_1 = 90°$, calculate the angle of misalignment 2 seconds after moving the control wheel.

(Part 3. I.E.E. 1958)

(8·25°)

7. Describe briefly the essentials of a simple remote-position-control servo-mechanism stabilized by direct velocity feed-back and show that its operation is characterized by a differential equation of the form

$$\frac{1}{\omega_n^2}\frac{d^2\theta_o}{dt^2}+T\frac{d\theta_o}{dt}+\theta_o = \theta_i$$

The position of a rotatable mass driven by an electric motor is controlled from a hand-wheel. The damping torques due to viscous friction and velocity feed-back, respectively, are equal. The moment of inertia of the moving parts referred to the mass is 100 kg-m^2 and the undamped natural frequency is 2·5 c/s. If the motion is critically damped, calculate the feed-back torque per unit angular velocity and the steady-state angular misalignment when the input angle is varied at the rate of 1 rad/s.

(Part 3. I.E.E. 1960)

(1571 N-m/rad/s; 0·127 radian)

8. The angular position of a flywheel which is driven by an electric motor is controlled from a hand-wheel employing a closed-loop automatic control system. The rotation of the flywheel is damped by viscous friction. Describe such a scheme with the aid of a block diagram and set up the differential equation of the system.

The moment of inertia of the moving parts referred to the velocity of the flywheel is 1000 kg-m^2 and the motor torque is 2 N-m per minute of misalignment. The total frictional torque, also referred to the flywheel velocity, is 2000 N-m/rad/s. At time $t = 0$, with the equipment at rest, the hand-wheel is set in motion with a constant angular velocity of 20 r.p.m. Derive the equation of the subsequent angular position of the flywheel in relation to time and make a sketch of this function.

(Part 3. I.E.E. 1959)

$[2{\cdot}1\,t - 0{\cdot}61 + 0{\cdot}87\mathrm{e}^{-t}\sin(2{\cdot}42\,t + 2{\cdot}36)]$

9. Draw a block diagram for an error-actuated automatic control system for the position control of a rotatable mass, the system being stabilized by output velocity feed-back. Describe briefly the purpose of the essential elements and deduce the differential equation of the system, neglecting friction.

A flywheel driven by an electric motor is automatically controlled to follow the movement of a hand-wheel. The inclusive moment of inertia of the flywheel is 100 kg-m^2 and the motor torque applied to it is 2500 N-m per radian of misalignment between the flywheel and the hand-wheel. A flywheel velocity of 1 rad/s produces a feed-back torque on the flywheel of 600 N-m. The hand-wheel is suddenly turned through 90° when the system is at rest. Derive an expression for the subsequent angular position of the flywheel in relation to time and sketch the form of this function.

(Part 3. I.E.E. 1962)

$[\pi/2(1 - \mathrm{e}^{-3t}(\cos 4t + \tfrac{3}{4}\sin 4t)]$

CHAPTER 5

System Analogues and Analogue Computing Units

*5.1 Introduction

In chapters 1 and 4 we have discussed the theory of a simple closed-loop control system and seen that many problems exist in the consideration of system elements. Imagine now that we are given the problem of designing a control system. How would we go about it? We must know the required output of the system, e.g. for a remote position control system the limit of output rotation and the range of output torque are specified. We must know the accuracy of control required. A rough sketch of the blocks comprising a possible system is made and system elements are chosen. Where do we go from here? Often experimental systems are made and altered by 'trial and error' until an operational system is obtained. The operational system can then be analysed and adjusted for the required performance. This method of designing a control system could be costly and time-consuming. It would be helpful if we could make a 'model' system with adjustable elements. Experiments on the model would enable us to design the required system without the lengthy and costly business of making experimental systems.

What do we mean by a system 'model'? A model motor car is generally a small copy of a motor car. An exact model would be one which has an engine and performs in exactly the same way as the real thing. The model need not have an internal combustion engine under the bonnet. Any type of engine might be used providing the resulting performance is the same. There will obviously be scale factors relating the size and performance of the car and its model. The model motor car is said to be *analogous* to the real motor car and therefore resembles it in such a way that its size and performance are in proportion to the size and performance of the real car. In as much as motor cars are controlled machines they are systems and their dynamic behaviour will have a mathematical description called the *system equation.* We are concerned with the behaviour of control systems, so we can go further and say that two systems will be considered *analogous* as long as their behaviour, defined by an equation, is identical.

Consider the following example:
Equation 4.26 is the system equation for a remote position control system

$$\Theta_i = \Theta_o\left(\frac{p^2}{\omega_n^2}+\frac{2\zeta}{\omega_n}p+1\right) \tag{4.26}$$

The equation of an electrical resistance-inductance-capacitance (*RLC*) series circuit with an applied voltage v and a circuit current of i can be found using Kirchhoff's second law.

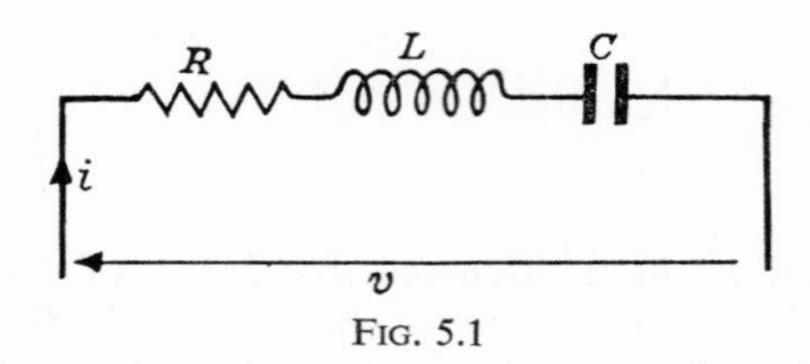

FIG. 5.1

Figure 5.1 shows the *RLC* circuit.
The equation is

$$v = iR+L\frac{d}{dt}i+\frac{1}{C}\int_0^t i\,dt \tag{5.1}$$

where the voltage v and the current i are functions of time t.

Changing equation 5.1 into operational form we have

$$V = IR+pLI+\frac{1}{pC}I \tag{5.2}$$

(p is the operation of differentiation with respect to time and $1/p$ is the operation of integration with respect to time).

V and I are functions of the p-operator. If we multiply both sides of equation 5.2 by pC we obtain

$$pCV = I(p^2LC+pCR+1) \tag{5.3}$$

Let $pCV = V_i$ the input and let I be the circuit output, therefore

$$V_i = I(p^2LC+pCR+1) \tag{5.4}$$

Equation 4.26, the system equation for a remote position control system, is analogous to equation 5.4 as long as $1/\omega_n^2$ is numerically equal to LC and $2\zeta/\omega_n$ is numerically equal to CR; V_i and I will represent Θ_i and Θ_o respectively.

We can say that as long as the coefficients of the *RLC* circuit equation can be made the same as the coefficients of the remote position control system equation, defined by equation 4.26, the two systems are analogous. Comparing equations 4.26 and 5.4,

$$I = V_i \left(\frac{1}{p^2LC + pCR + 1} \right)$$

output = input × 'same' transfer function

$$K\Theta_o = \Theta_i \left(\frac{1}{p^2/\omega_n^2 + (2\zeta/\omega_n)p + 1} \right)$$

numerically,

$LC = 1/\omega_n^2$

$CR = 2\zeta/\omega_n$

K is a constant

As the two systems are analogous and K is a scale factor, a suitable *RLC* circuit can be used to *compute* the behaviour of I with any value of V. Therefore, the circuit can be used to *compute* the behaviour of Θ_o with any value Θ_i. The circuit computes the behaviour of a remote position control system damped by viscous friction. It can be called an *analogue computer.*

The circuit described is not a practical form of analogue computer as will become clear in the following sections.

5A Introduction to the analogue computer

We have seen how a series *RLC* circuit can be used as an analogue computer to compute the output of a remote position control system for any input. The converse is true; the remote position control system could compute the output of a series *RLC* circuit. Any system that has the same mathematical description as another can be used as an analogue computer.

All linear systems can be described by their transfer functions. A general transfer function may be described in the following form

$$\frac{\text{output}}{\text{input}} = \frac{a_0 + a_1p + a_2p^2 + a_3p^3 + \ldots + a_np^n}{b_0 + b_1p + b_2p^2 + b_3p^3 + \ldots + b_mp^m} \tag{5.5}$$

where, for the systems we shall consider, $m > n$, a_0, a_1, a_2 to a_n and b_0, b_1, b_2 to b_m are constant coefficients and p is an operator. For an analogue computer to compute the behaviour of any linear system it seems likely that units which can perform mathematical functions must be included. Although it has been pointed out that any system could be used as an analogue computer, electrical systems are generally used. We shall consider electrical systems that can perform the operations of addition, differentiation and integration.

Summary

(*i*) Two systems are considered to be analogous if their behaviour is identically in proportion by mathematical definition.

(*ii*) An analogue computer is a system which can compute the behaviour of another system or other systems.

5B(i) Analogue computer units

In this section we shall show the development of electrical analogue computing units, but without going too far into the electronic circuitry. A familiarity with electronic amplifiers will, however, be assumed.

(*i*) *Passive analogue computing circuits*

(*a*) *Passive adding circuit*

A passive circuit is one which does not have active elements such as valves, transistors, etc.

An adding circuit has an *output* which is the *sum of a number of inputs.* The circuit shown in figure 5.2 is an adding circuit.

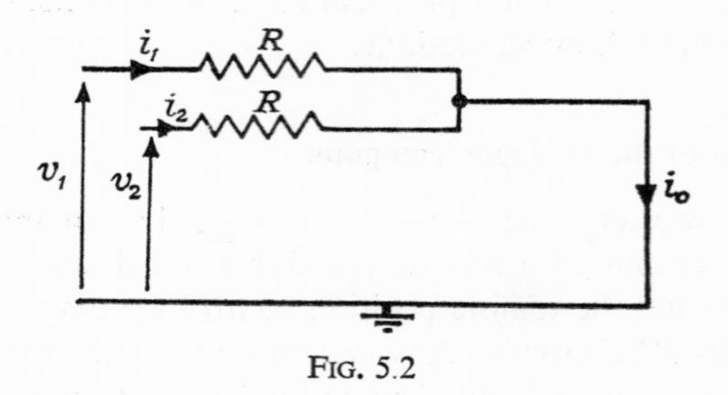

FIG. 5.2

Figure 5.2. The inputs to the circuit are the voltages v_1 and v_2, the output of the circuit is a current i_o. The input currents are given by $i_1 = v_1/R$ and $i_2 = v_2/R$. The output current, by Kirchhoff's first law, is given by

$$i_o = i_1+i_2$$

therefore

$$i_o = v_1/R+v_2/R$$

$$i_o = \frac{1}{R}(v_1+v_2),$$

$$i_o = \text{constant}\,(v_1+v_2) \tag{5.6}$$

$$\text{output} = \text{constant} \times \text{sum of the inputs}$$

Equation 5.6 shows that the circuit in figure 5.2 can be used as an adding circuit. The constant is a known scale factor. The output of the circuit is a current and the input a voltage. This is undesirable if we wish to connect two circuits together, because we would require the output and the input to have the same dimensions. We can, however, obtain an output voltage if the circuit is arranged as in figure 5.3.

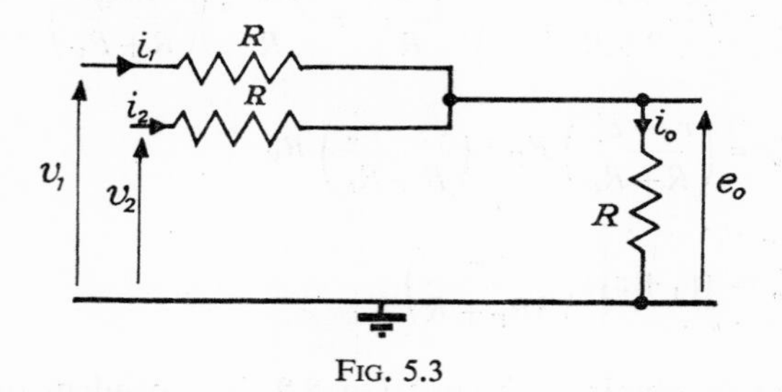

FIG. 5.3

Figure 5.3. The inputs to the circuit are the voltages v_1 and v_2; the output of the circuit is a voltage e_o. The input currents are given by

$$i_1 = \frac{v_1 - e_o}{R} \quad \text{and} \quad i_2 = \frac{v_2 - e_o}{R}$$

$$i_o = i_1 + i_2,$$

therefore

$$i_o = \frac{v_1 - e_o}{R} + \frac{v_2 - e_o}{R} \tag{5.7}$$

substituting $e_o = i_o R$ into equation 5.7, we obtain

$$\frac{e_o}{R} = \frac{v_1}{R} - \frac{e_o}{R} + \frac{v_2}{R} - \frac{e_o}{R}$$

$$e_o = \tfrac{1}{3}(v_1 + v_2) \tag{5.8}$$

$$\text{output} = \text{constant} \times \text{sum of the inputs}$$

Equation 5.8 shows that the circuit in figure 5.3 can be used as an adding circuit. The disadvantage with this circuit is that when its output is connected to another circuit the effective value of the output voltage changes. This can be demonstrated by considering the effect of connecting a voltmeter of resistance R_V across X and Y. This is shown in figure 5.4.

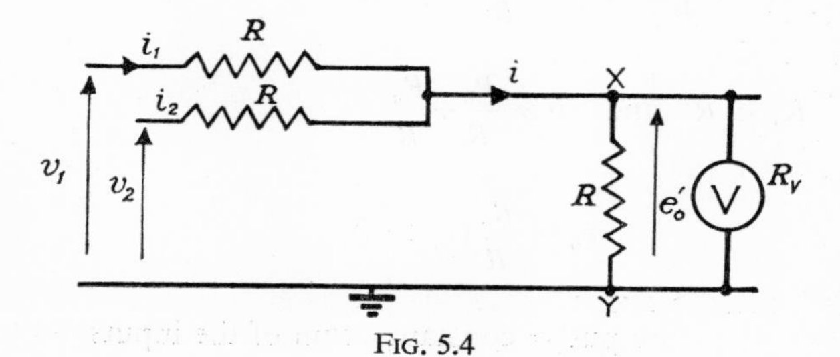

FIG. 5.4

We will find the new value of output voltage e'_o,

$$e'_o = i\left(\frac{RR_V}{R+R_V}\right) = \left(\frac{v_1-e'_o}{R}+\frac{v_2-e'_o}{R}\right)\left(\frac{RR_V}{R+R_V}\right)$$

$$e'_o = \left(\frac{v_1-e'_o}{R+R_V}\right)R_V+\left(\frac{v_2-e'_o}{R+R_V}\right)R_V$$

$$e'_o = (v_1+v_2)\left(\frac{R_V}{3R_V+R}\right) \tag{5.9}$$

The value of e'_o obtained in equation 5.9, is dependent on R_V, the load impedance. Therefore the circuit shown in figure 5.4 can only be used as an adding circuit for one value of load impedance. One way of connecting a load across X and Y such that it has little effect on the circuit's operation is to make the impedance R_o much smaller than R. This is shown as follows:

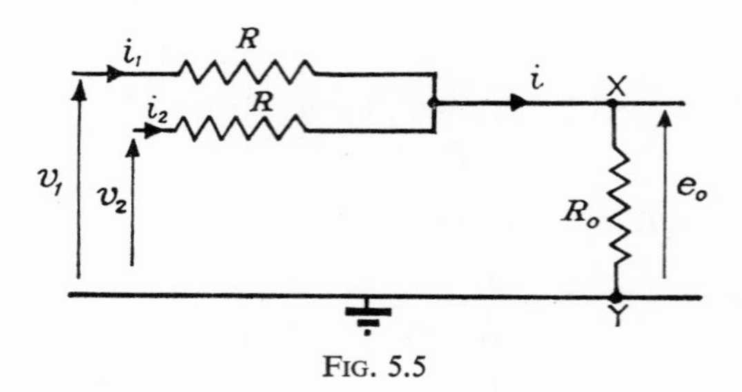

FIG. 5.5

Figure 5.5. In this circuit the resistance R_o is much smaller than the resistance R and the output voltage e_o is given by

$$e_o = iR_o = \left(\frac{v_1-e_o}{R}+\frac{v_2-e_o}{R}\right)R_o$$

$$e_o\left(1+\frac{R_o}{R}+\frac{R_o}{R}\right) = \frac{R_o}{R}(v_1+v_2)$$

as $$R_o \ll R \quad \text{and} \quad 1 \gg \frac{R_o}{R}+\frac{R_o}{R}$$

$$e_o \simeq \frac{R_o}{R}(v_1+v_2)$$

$$\text{output} \simeq \text{constant} \times \text{sum of the inputs}$$

Therefore we can see that as R_o is small the connection of an impedance across X and Y, of the order of R, will have little effect on the circuit operation. The scale factor R_o/R is very small. This means that the output voltage will be very small compared with the input voltages.

(*b*) *Passive differentiating circuit*

A differentiating circuit has an *output* which is the *input differentiated.* The circuit shown in figure 5.6 is a differentiating circuit.

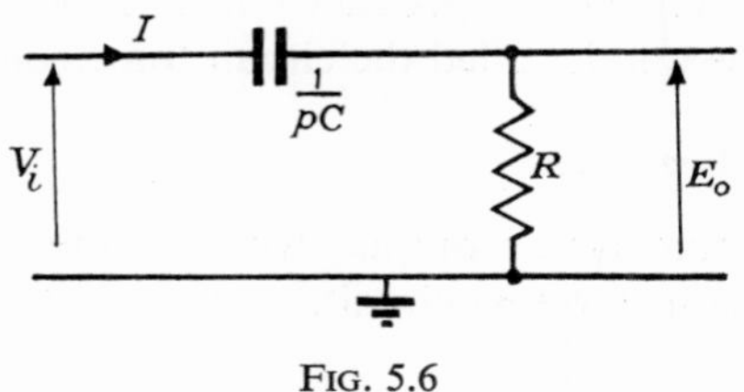

FIG. 5.6

Figure 5.6. The capacitor in the circuit is specified as $1/pC$. This is the operational impedance of a capacitor of capacitance C. The voltage and current relationship of a capacitor of capacitance C is

$$V = \frac{1}{C}\int_0^t i\,dt \tag{5.10}$$

The operational way of writing integration with respect to time is $1/p$, therefore equation 5.10 can be written as follows

$$V = \frac{1}{C}\,\frac{1}{p}\,I$$

$$V = \frac{1}{pC}\,I \tag{5.11}$$

In equation 5.11 the term $1/pC$ operates on the current I to give the voltage V hence, $1/pC$ having the dimension of resistance, is called the *impedance operator* of a capacitor (compare with Ohm's law).

The input to the circuit is a voltage V_i and the output is a voltage E_o. The input current I is given by

$$I = \frac{V_i - E_o}{1/pC} = \frac{E_o}{R}$$

therefore

$$p(V_i - E_o) = \frac{1}{CR}\,E_o \tag{5.12}$$

Equation 5.12 indicates that if E_o is very small compared with V_i, we obtain

$$pV_i \simeq \frac{1}{CR} E_o \tag{5.13}$$

input differentiated $\simeq$ constant $\times$ output

This is the desired result as p is the operator notation for d/dt and $1/CR$ is a constant scale factor. For E_o to be small compared with V_i we can see from equations 5.12 and 5.13 that the product CR must be small. (CR will have the dimension of time and is called the circuit 'time constant'.)

(*c*) *Passive integrating circuit*

An integrating circuit has an output, which is the *input integrated.* The following circuit is an integrating circuit.

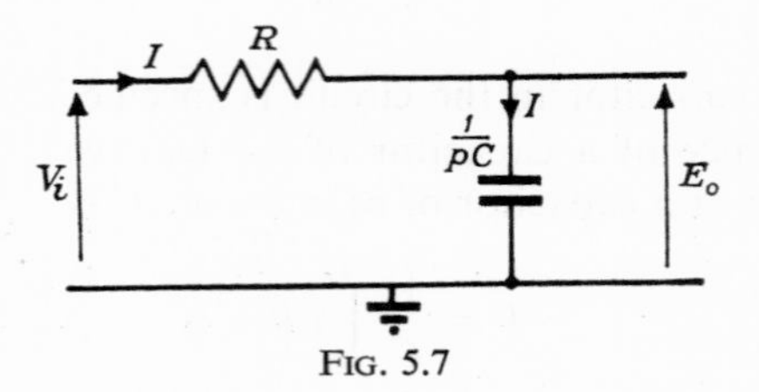

FIG. 5.7

Figure 5.7. The input to the circuit is a voltage V_i and the output is a voltage E_o. The input current is given by

$$I = \frac{V_i - E_o}{R} = \frac{E_o}{1/pC}$$

therefore

$$\frac{1}{p}(V_i - E_o) = CRE_o \tag{5.14}$$

Equation 5.14 indicates that if E_o is very small compared with V_i we obtain

$$\frac{1}{p} V_i = CRE_o \tag{5.15}$$

input integrated $=$ constant $\times$ output

This is the desired result as 1/p is the operator notation for $\int_0^t f(t)\,dt$ and CR is a constant scale factor. For E_o to be small compared with V_i we can see, from equations 5.14 and 5.15, that the product CR must be large.

Summary

(*a*) *Passive addition*

Conditions for addition: $R_o \ll R$, hence $e_o \ll (v_1+v_2)$

output $\simeq$ constant $\times$ sum of the inputs

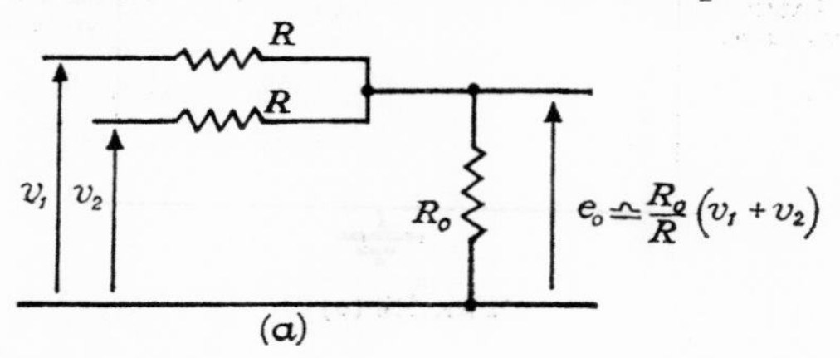

(*b*) *Passive differentiation*

Conditions for differentiation: $E_o \ll V_i$, hence CR is small.

output $\simeq$ constant $\times$ input differentiated

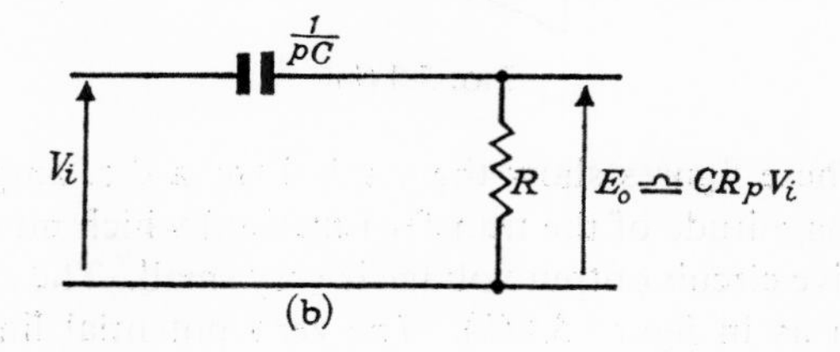

(*c*) *Passive integration*

Conditions for integration: $E_o \ll V_i$, hence CR is large.

output $\simeq$ constant $\times$ input integrated

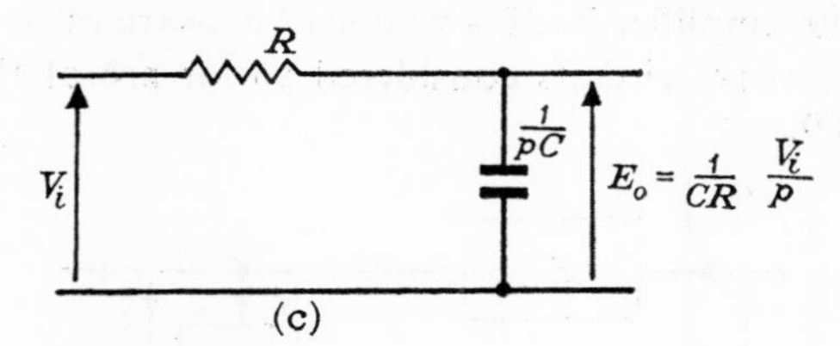

Conclusions

The passive circuits in all three cases do the required operation, but the outputs obtained are very small compared with the inputs. This is undesirable. It is obvious that an electronic amplifier could amplify the outputs. The input functions (V_i and (v_1+v_2)) could be of any value, therefore a directly coupled amplifier must be used, i.e. an amplifier which will operate at frequencies down to and including 'zero frequency' (direct current).

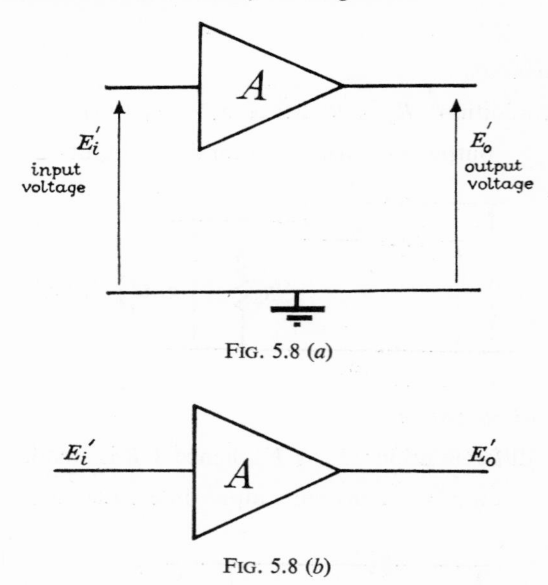

FIG. 5.8 (*a*)

FIG. 5.8 (*b*)

Figure 5.8. These figures show the symbol for a d.c. amplifier. A is the amplifier gain (magnitude of the transfer function) which must be very large, because the passive circuit output voltage is very small. The amplifier symbol is usually drawn as in figure 5.8(*b*). The zero potential line is omitted, it being assumed that all potentials are measured with respect to zero potential.

(*ii*) *The d.c. amplifier with feed-back*

Before discussing the use of the d.c. amplifier with analogue computing units we shall discuss the amplifier itself—we must be aware of its advantages and limitations. The passive circuits considered so far are of the general form shown in figure 5.9.

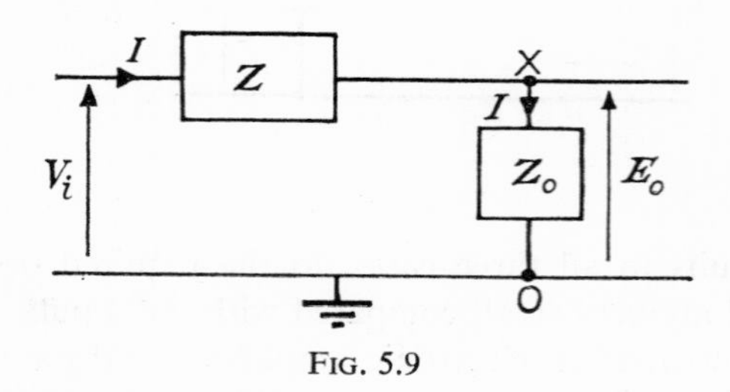

FIG. 5.9

Figure 5.9. The output voltage E_o is small compared with the input voltage, and to obtain a useful output a d.c. amplifier is used.

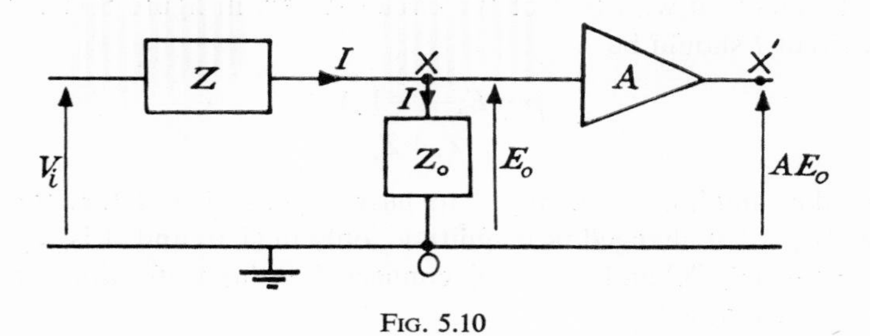

FIG. 5.10

Figure 5.10. The amplifier is connected to the output of the circuit in figure 5.9. It is not considered to take any appreciable current, i.e. it does not load the passive circuit.

The output voltage is now AE_o. A is a very large value to make the product AE_o a useful voltage. The circuit shown in figure 5.10 is not practical because, at very high gains, d.c. amplifiers tend to become unstable, and drift, and therefore a particular value of A cannot be held constant. This problem is overcome by rearranging the circuit shown in figure 5.10.

We shall consider the rearrangement in a series of circuits, so as to obtain a full understanding of it. First of all a break in the circuit is made at the junction of the impedance Z_o and the zero potential line. X must be at a potential of E_o and O must be at a potential of zero for the circuit to be unchanged in operation. An impedance Z_F is connected to the free end of Z_o, the other end of the impedance Z_F being connected to a supply of voltage $-V_o$, such that X is still at the potential E_o and O at zero potential.

The circuit will now be:

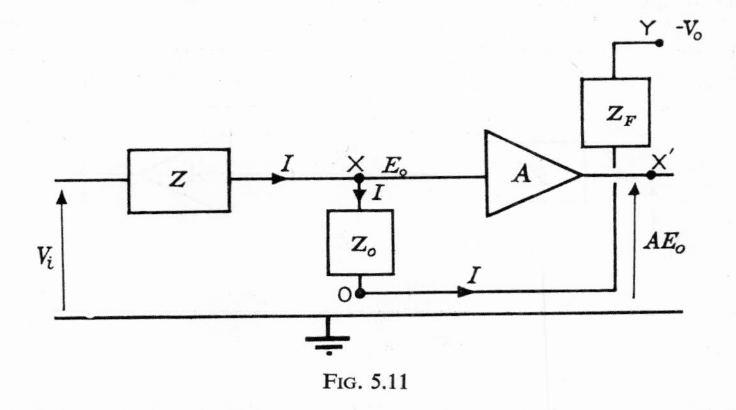

FIG. 5.11

Figure 5.11. The condition for the operation of the circuit shown in figure

5.10 to be identical with that of the circuit shown in figure 5.11 is that the input current I should be

$$I = \frac{E_o-(-V_o)}{Z_o+Z_F}$$

If the d.c. amplifier is arranged to have a gain of $-A$ (i.e. the output voltage AE_o is not changed in magnitude, only in sign) and A is adjusted so that $V_o = -AE_o$, X′ and Y can be connected without affecting the circuit. The circuit will now be:

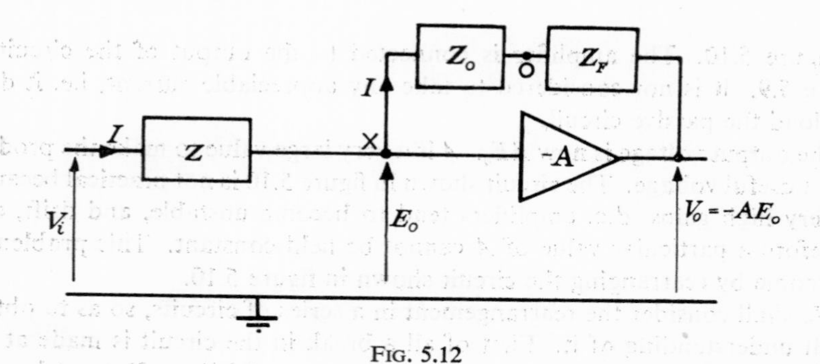

FIG. 5.12

Figure 5.12. The sign of the output is taken care of by marking the amplifier gain negative.

If A is made very large E_o can be extremely small, and Z_F is adjusted so that it is very much larger than Z_o. The impedance Z_o can, under these conditions, be neglected compared with the impedance Z_F. The circuit is now as in figure 5.13, Z_o being neglected.

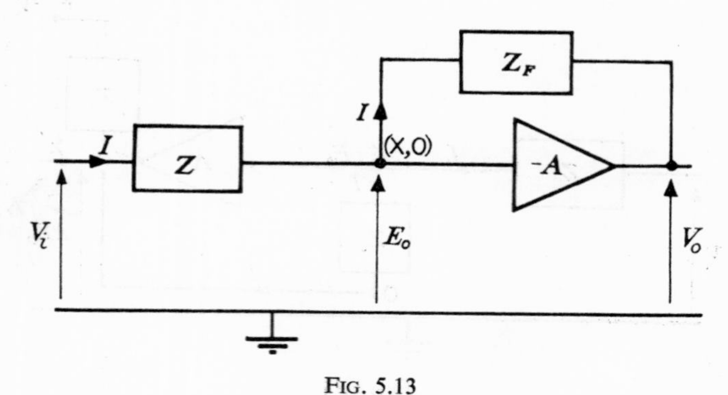

FIG. 5.13

Figure 5.13. Points X and O are coincident. (It is important to note that although, theoretically, point O is at zero potential, for the output V_o to exist point (X, O) must be at the extremely small potential of E_o.) The impedance Z_F is a *feed-back* device and the advantages of this arrangement of the circuit will become clearer if we examine the overall circuit transfer function (in this case the transfer function is often called the *gain*) and the driving-point impedance at X.

Overall transfer function

Let all the potentials and impedances considered be functions of the operator p. The electronic amplifier will be considered not to take any appreciable current from the circuit. From figure 5.13 the input current I is given by

$$\frac{V_i - E_o}{Z} = I$$

and

$$\frac{E_o - V_o}{Z_F} = I \qquad (5.16)$$

therefore

$$\frac{V_i - E_o}{Z} = \frac{E_o - V_o}{Z_F} \qquad (5.17)$$

In equation 5.17, by definition, E_o will be very small compared with the input voltage V_i and the output voltage V_o and therefore can be neglected.

Hence equation 5.16 becomes

$$\frac{V_i}{Z} = -\frac{V_o}{Z_F}$$

therefore

$$\frac{V_o}{V_i} = \frac{\text{output}}{\text{input}} = -\frac{Z_F}{Z} \qquad (5.18)$$

Equation 5.18 gives the overall transfer function (gain). It is negative and depends only on the impedances Z_F and Z; therefore the transfer function must be constant. With a constant transfer function the amplifier will be stable. As the impedances Z_F and Z can be operational impedances, this arrangement of a d.c. amplifier can itself be operational and is called an *operational amplifier*.

Driving-point impedance at X

This is defined as the ratio of the potential at X, with respect to zero potential, to the current flowing into X. The driving-point impedance at X is E_o/I. Equation 5.16 is

$$\frac{E_o - V_o}{Z_F} = I$$

but we know that $V_o = -AE_o$, therefore substituting $-AE_o$ for V_o in equation 5.16, we obtain

$$\frac{E_o + AE_o}{Z_F} = I$$

$$\frac{E_o(1+A)}{Z_F} = I$$

therefore $$\frac{E_o}{I} = \text{driving-point impedance at X} = \frac{Z_F}{1+A} \tag{5.19}$$

Equation 5.19 is the driving-point impedance at X. This equation enables us to draw the general equivalent circuit for the operational amplifier as follows:

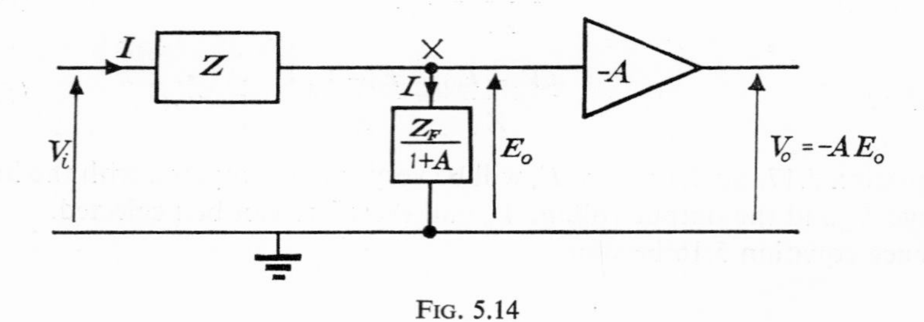

FIG. 5.14

Figure 5.14. This equivalent circuit will be used in the discussion on active analogue computing units.

Summary

(*i*) A d.c. amplifier with a feed-back impedance is stable and has an unvarying overall transfer function (gain) dependent on the ratio of two operational impedances ($-Z_F/Z$).

(*ii*) The general equivalent circuit of the *operational amplifier* is as shown in figure 5.14 and the conditions for a constant transfer function are that:

E_o tends to zero
A tends to infinity

5B(ii) Active analogue computing units

Active analogue computing circuits can be made incorporating the operational amplifier.

(*a*) *Active adding unit*

The circuit of an active adding unit is shown in figure 5.15(*a*) and its equivalent circuit is shown in figure 5.15(*b*).

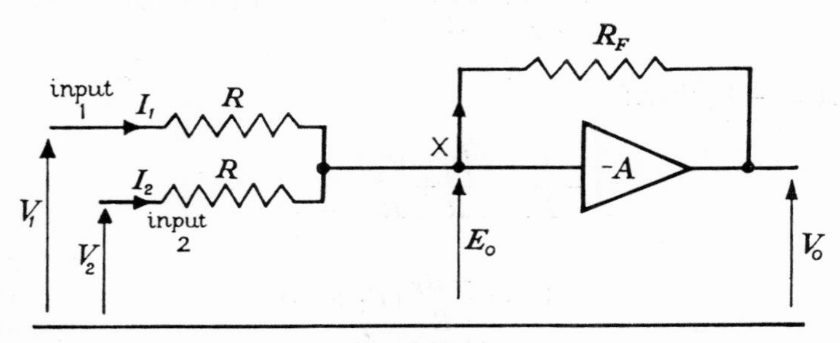

FIG. 5.15 (*a*)

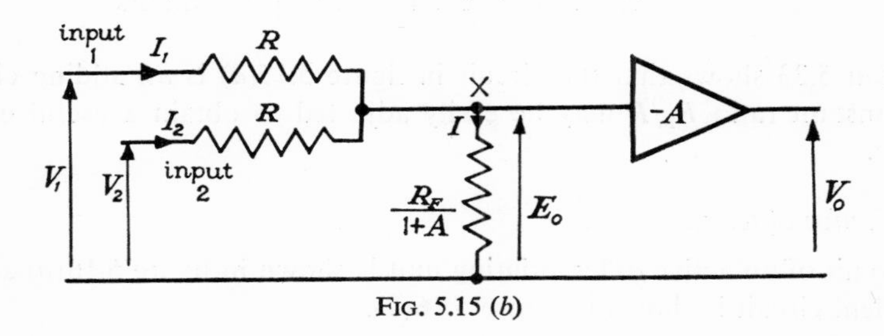

FIG. 5.15 (*b*)

Figure 5.15(*b*) has two input currents of I_1 and I_2. Input current I_1 is given by

$$I_1 = \frac{V_1 - E_o}{R}$$

Input current I_2 is given by

$$I_2 = \frac{V_2 - E_o}{R}$$

The current I at X is given by

$$I = \frac{V_1 - E_o}{R} + \frac{V_2 - E_o}{R} \tag{5.20}$$

where

$$I = E_o \bigg/ \frac{R_F}{1+A}$$

Substituting this value of current I into equation 5.20 we obtain

$$\frac{E_o(1+A)}{R_F} = \frac{V_1 - E_o}{R} + \frac{V_2 - E_o}{R} \tag{5.21}$$

We know, by definition, that voltage E_o tends to zero and therefore equation 5.21 becomes

$$\frac{E_o A}{R_F} = \frac{V_1}{R} + \frac{V_2}{R} \qquad (A+1 \simeq A) \tag{5.22}$$

but $E_o A = -V_o$ therefore

$$-\frac{V_o}{R_F} = \frac{V_1}{R} + \frac{V_2}{R}$$

$$V_o = \frac{-R_F}{R}(V_1 + V_2) \tag{5.23}$$

output = constant × sum of the inputs

Equation 5.23 shows that the circuit in figure 5.15(*a*) is an adding circuit. The constant ratio R_F/R may be easily adjusted to obtain a useful output voltage.

(*b*) *Active differentiating unit*

The circuit of an active differentiating unit is shown in figure 5.16(*a*) and its equivalent circuit is shown in figure 5.16(*b*).

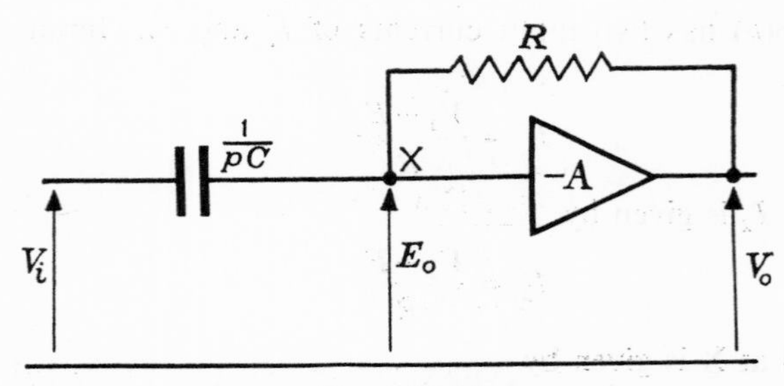

FIG. 5.16 (*a*)

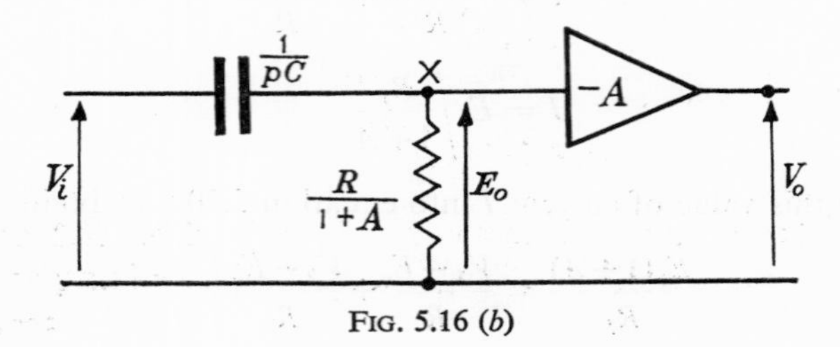

FIG. 5.16 (*b*)

Figure 5.16(*b*). The voltage E_o is given by

$$E_o = \frac{V_i\left(\dfrac{R}{1+A}\right)}{1/pC+\left(\dfrac{R}{1+A}\right)}$$

$$E_o = \frac{\dfrac{CR}{1+A}pV_i}{1+\dfrac{pCR}{1+A}} \tag{5.24}$$

If we multiply both sides of equation 5.24 by A and put $V_o = -AE_o$ we obtain

$$V_o = \frac{-\left(\dfrac{A}{1+A}\right)CRpV_i}{1+\dfrac{pCR}{1+A}} \tag{5.25}$$

In equation 5.25, A tends to infinity, therefore the ratio $A/(1+A)$ is approximately unity and the ratio $pCR/(1+A)$ tends to zero. Equation 5.25 becomes

$$V_o = -CRpV_i \tag{5.26}$$

output = constant × input differentiated

Equation 5.26 shows that the circuit in figure 5.16(*a*) can be used as a differentiating circuit. The time constant CR may be easily adjusted to obtain a useful output voltage V_o.

In practice differentiating units are not used, as differentiation can easily lead to operational amplifier overloading. As an example consider an operational amplifier that will overload with an input signal greater than 100 volts. Let the input signal be sinusoidal and 10 sin 500t volts. The output will be the input differentiated, i.e. 5000 cos 500t. The output signal has a peak value of 5000 volts, which is fifty times the maximum allowable input signal! As thermal noise contains relatively high frequencies it can often swamp an output signal when differentiation is attempted. As differentiation can so easily lead to overloading, the details of differentiating circuits for practical analogue computing will not be considered any further.

(*c*) *Active integrating unit*

The circuit of an active integrating unit is shown in figure 5.17(*a*) and its equivalent circuit is shown in figure 5.17(*b*).

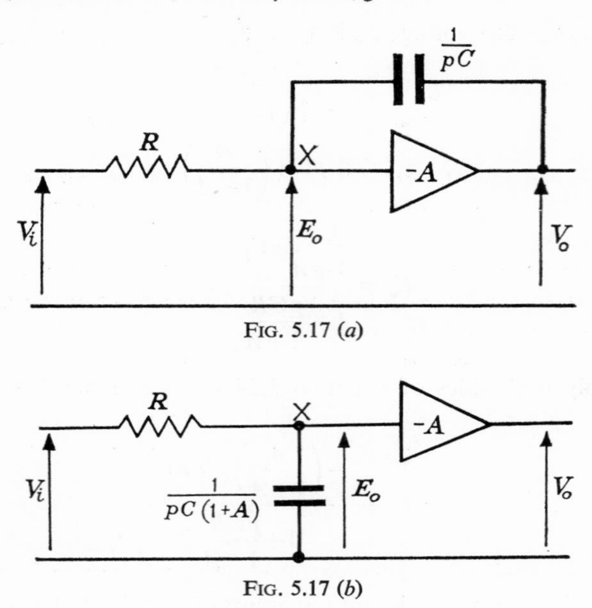

FIG. 5.17 (*a*)

FIG. 5.17 (*b*)

Figure 5.17(*b*). The voltage V_o is given by

$$E_o = \frac{V_i \dfrac{1}{pC(1+A)}}{R + \dfrac{1}{pC(1+A)}}$$

$$E_o = \frac{\dfrac{V_i}{CR} \dfrac{1}{p} \dfrac{1}{(1+A)}}{1 + \dfrac{1}{pCR(1+A)}} \tag{5.27}$$

If we multiply both sides of equation 5.27 by A and put $V_o = -AE_o$ we obtain

$$V_o = \frac{-\left(\dfrac{A}{1+A}\right) \dfrac{1}{CR} \dfrac{V_i}{p}}{1 + \dfrac{1}{pCR(1+A)}} \tag{5.28}$$

In equation 5.28, A by definition tends to infinity, the ratio $A/(1+A)$ is approximately unity and the ratio $1/pCR(1+A)$ tends to zero. Equation 5.28 becomes

$$V_o = -\frac{1}{CR}\frac{1}{p}V_i \tag{5.29}$$

output = constant × input integrated

Equation 5.29 shows that the circuit in figure 5.17(*a*) can be used as an integrating circuit. The time constant *CR* may be easily adjusted to give a useful output voltage V_o.

It should be remembered that operational amplifiers are usually drawn omitting the zero potential line, it being understood that all potentials marked are measured with respect to zero potential.

Figure 5.18 shows a general operational amplifier.

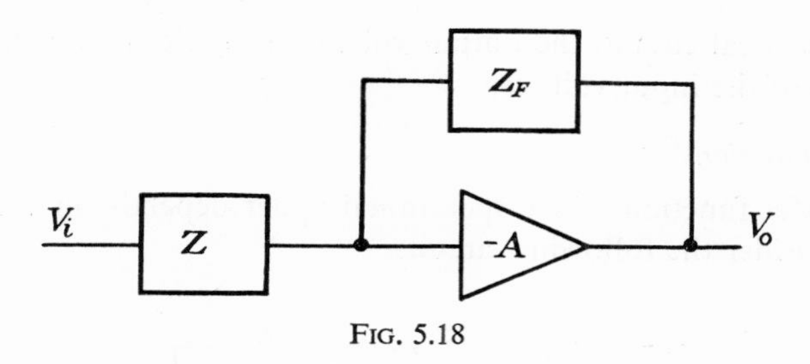

FIG. 5.18

Summary and conclusions

The operational adder

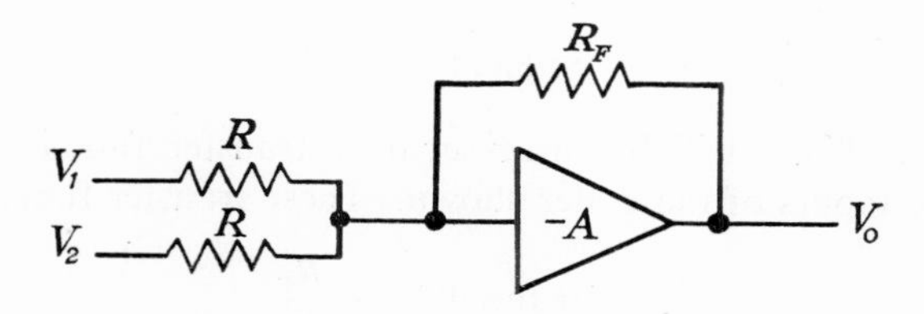

The output voltage V_o is a function of the sum of the input voltages $V_1 + V_2$

$$V_o = -\frac{R_F}{R}(V_1 + V_2)$$

This is a practical circuit, the output voltage being negative with respect to the sum of the input voltages.

The operational differentiator

The operational differentiator is not a practical circuit due to likely amplifier overloading.

The operational integrator

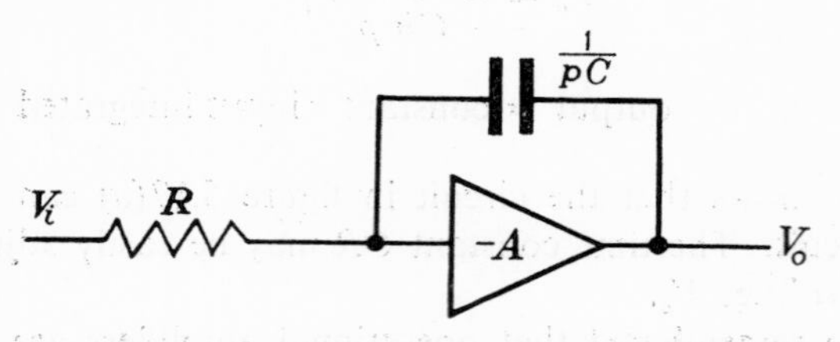

The output voltage V_o is a function of the integral of the input voltage.

$$V_o = -\frac{1}{CR}\frac{V_i}{p}$$

This is a practical circuit, the output voltage being negative with respect to the integral of the input voltage.

Transfer functions

The transfer function of an operational adder depends on the individual inputs. Consider the following circuit,

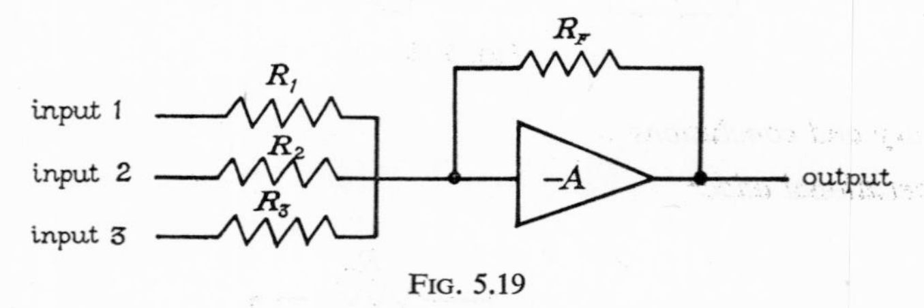

FIG. 5.19

Figure 5.19. There will be three separate transfer functions associated with the three inputs of the adder shown. These transfer functions will be:

(*i*) for input 1 $\quad -\dfrac{R_F}{R_1}$

(*ii*) for input 2 $\quad -\dfrac{R_F}{R_2}$

(*iii*) for input 3 $\quad -\dfrac{R_F}{R_3}$

The transfer function of an operational differentiator will not be considered, as the circuit is not of any real practical value.

The transfer function of an operational integrator is $\dfrac{-1/pC}{R}$. Let the time constant CR be T. The transfer function of the operational integrator is now $-1/pT$.

The inverter

The operational amplifiers discussed all have outputs which are negative compared with the required output function. This is one reason why an inverting device is useful.

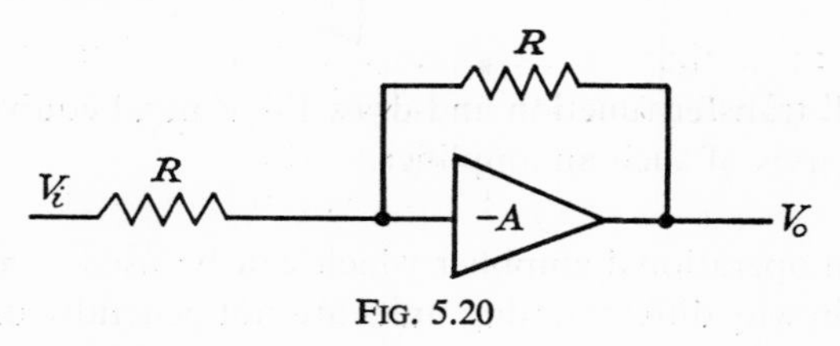

FIG. 5.20

Figure 5.20 shows an inverter.
The transfer function of the amplifier is unity and negative ($-R/R$); therefore the output voltage V_o will be the input voltage V_i inverted.

The divider

Although not a conclusion of this section, it is necessary to discuss simple division before we examine the use of the practical operational units. Division by a constant is obtained by using a potential divider, shown in figure 5.21(*a*). The symbol for the divider is shown in figure 5.21(*b*).

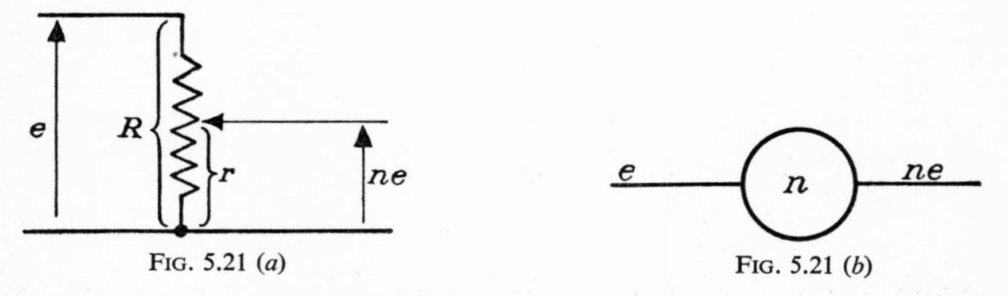

FIG. 5.21 (*a*) FIG. 5.21 (*b*)

The transfer function n of the divider is r/R.

In the next chapter we shall see how we can use the analogue units we have developed to examine the behaviour of systems.

5C Examples

1. Explain what is meant by a system analogue giving brief examples.

2. Discuss the drawbacks of employing passive differentiating and integrating circuits.

3. Explain why feed-back is employed in a high-gain d.c. amplifier.
The figure shows an operational amplifier.

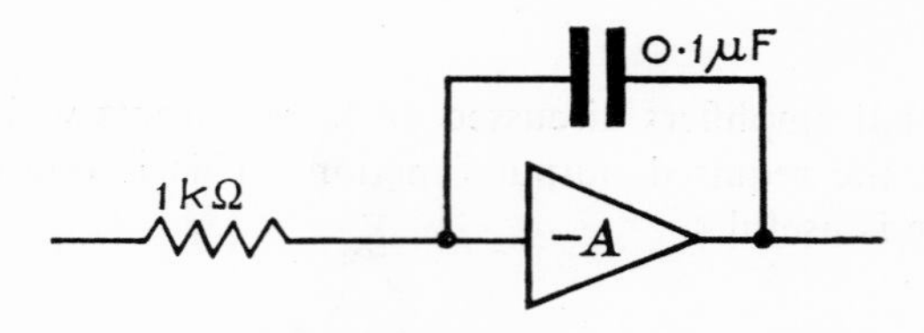

Find the overall transfer function and draw the general equivalent circuit.
Describe the uses of such an amplifier.

$(10^4/p)$

4. Describe an operational amplifier which can be used as a differentiating unit, and explain why differentiating units are not generally used in practice.

CHAPTER 6

An Introduction to the Practical Use of an Analogue Computer

6A Introduction

Basically an analogue computer consists of a number of d.c. amplifiers arranged so that they can easily be connected up as operational amplifiers. For this purpose a *patch panel*, i.e. a panel with the relevant input, feed-back and output connections, is used. Each amplifier can be connected up in the desired operational fashion. Accurate components are available on the panel as input and feed-back elements.

To illustrate simple analogue computing methods and to show some of the problems involved, we shall use the operational units developed in chapter 5 to compute the behaviour of two systems. It should be stressed that an equation defining the system whose behaviour is to be computed must be known.

6B Example 1 of the use of an analogue computer

Undamped remote position control system

We shall examine how analogue computing units can be used to compute the behaviour of the output of a simple undamped remote position control system. In chapter 4 we arrived at a remote position control system equation with damping, equation 4.26,

$$\left(\frac{p^2}{\omega_n^2}+\frac{2\zeta p}{\omega_n}+1\right)\Theta_o = \Theta_i \tag{4.26}$$

For an undamped system the damping ratio ζ is zero and equation 4.26 becomes

$$\left(\frac{p^2}{\omega_n^2}+1\right)\Theta_o = \Theta_i \tag{6.1}$$

It has been suggested that the system impulse response is a useful solution (chapter 2, section 2.4). To obtain the impulse response the input in operational form is unity. With the input unity equation 6.1 becomes

$$\Theta_o = \frac{\omega_n^2}{p^2+\omega_n^2} \tag{6.2}$$

Changing equation 6.2 into a function of time t using the table of operator notation (chapter 2, p. 23) the solution will be of the form

$$\theta_o = \omega_n \sin \omega_n t \tag{6.3}$$

As the solution is to be computed its form is generally not known, or obvious, but for the consideration of this first example it is helpful. To enable analogue computing units to solve equation 6.2 we rearrange it in the form

$$\frac{p^2\Theta_o}{\omega_n^2}+\Theta_o = 1 \tag{6.4}$$

The procedure of setting up this equation for analogue computation will become clear as we follow the steps usually taken. Equation 6.4 is rewritten with the term containing the highest order of p on the left-hand side and all other terms on the right-hand side.

$$\frac{p^2\Theta_o}{\omega_n^2} = 1-\Theta_o \tag{6.5}$$

The equation is of the form

$$\frac{p^2E}{\omega_n^2} = 1-E \tag{6.6}$$

where the voltage E is numerically equal to the position Θ_o and ω_n^2 is a constant. Let the input to an integrator of transfer function $-1/(pT_1)$ be p^2E/ω_n^2. Therefore the output voltage will be

$$-\frac{p^2E}{\omega_n^2}\frac{1}{pT_1} = -\frac{pE}{\omega_n^2T_1}$$

as shown in figure 6.1.

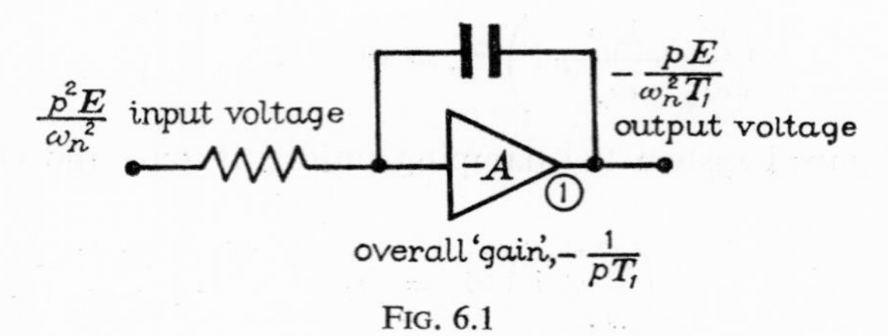

FIG. 6.1

Figure 6.1. The amplifier is labelled 1 and its output voltage is now applied to another integrator of 'gain' $-1/(pT_2)$. The output voltage of the second integrator, labelled 2, will be

$$-\frac{pE}{\omega_n^2 T_1}\left(-\frac{1}{pT_2}\right) = \frac{E}{\omega_n^2 T_1 T_2}$$

as shown in figure 6.2.

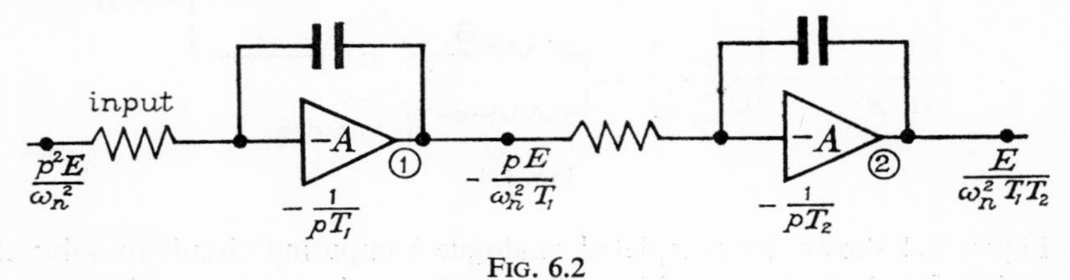

FIG. 6.2

Figure 6.2. The first amplifier in this figure has an input voltage of $(p^2E)/\omega_n^2$. Equation 6.6 shows that the term $(p^2E)/\omega_n^2$ must be the sum of an impulse voltage and a voltage $-E$. A third amplifier is now connected to the output of amplifier 2.

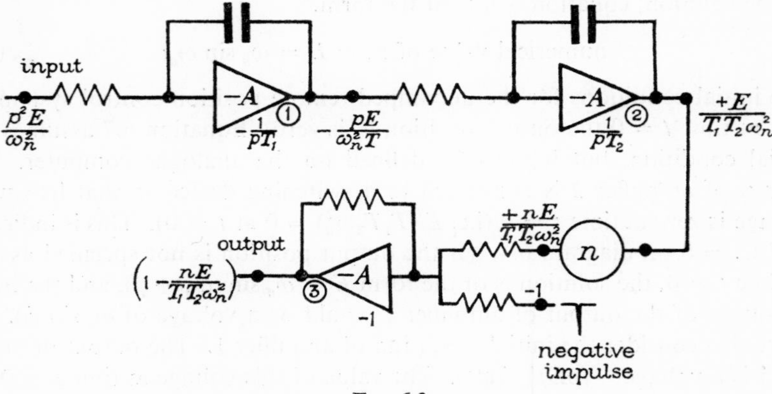

FIG. 6.3

Figure 6.3. In this figure an adder (amplifier 3) is connected to the output of amplifier 2. The transfer function for each input is arranged to be unity (-1). A divider is connected between amplifiers 2 and 3. The output of amplifier 3 is the sum of an impulse voltage and a voltage $-(nE)/(\omega_n^2 T_1 T_2)$. If $n/(T_1 T_2)$ is arranged to be the constant ω_n^2 the output of amplifier 3 becomes the voltage $(1-E)$. This is equal to the right-hand side of equation 6.6. If the output of amplifier 3 and amplifier 1 are connected together point A will be common and the required condition of $(p^2E)/\omega_n^2 = 1-E$ is obtained (figure 6.4).

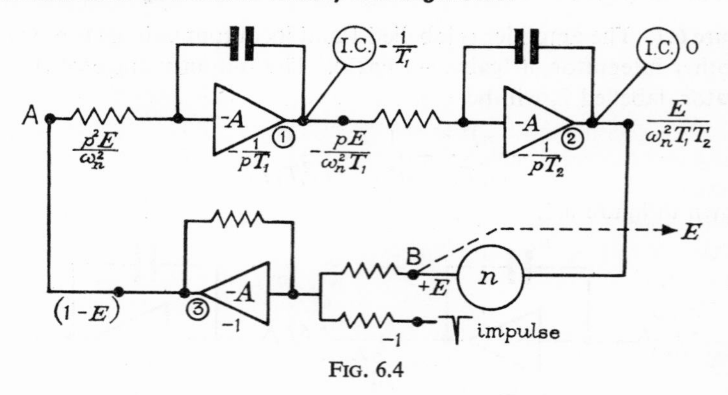

FIG. 6.4

Figure 6.4 shows the completed analogue computing circuit to solve the equation for the output position of an undamped remote position control system. The voltage at B is $(nE)/(T_1T_2\omega_n^2)$ and if n is adjusted so that $n = T_1T_2\omega_n^2$, the voltage E is obtained at B. Voltage E is numerically equal to the output position Θ_o and, if applied to a cathode ray oscilloscope, or an XY plotter, will give a graphical representation of the solution.

The solution, equation 6.3, is of the form,

$$\text{numerical value of } \theta_o = E = \omega_n \sin \omega_n t \tag{6.7}$$

The initial conditions for the undamped remote position control system are that at time $t = 0$ the output position θ_o is zero. Equation 6.7 assumes this initial condition, but it must be defined on the analogue computer. The output of amplifier 2 is connected to a switching device so that its output voltage is zero at time $t = 0$ (i.e. $E/(T_1T_2\omega_n^2) = 0$ at $t = 0$). This is indicated by I.C. = 0 on diagram 6.4. (If the output position is not specified as zero at time $t = 0$, the solution is of the form $E = \omega_n \sin(\omega_n t + \phi)$, and the initial condition of the output of amplifier 2 would be a voltage of $\omega_n \sin \phi$). We must also consider the initial condition of amplifier 1. The output of amplifier 1 is a voltage $-(pE)/(T_1\omega_n^2)$. The value of this voltage at time $t = 0$ can be found by differentiating ($p = d/dt$) equation 6.7, multiplying it by $-1/(T_1\omega_n^2)$ and substituting time $t = 0$, giving

$$-\frac{pE}{T_1\omega_n^2} = -\left[\frac{\omega_n^2 \cos \omega_n t}{T_1\omega_n^2}\right]_{t=0}$$

$$-\frac{pE}{T_1\omega_n^2} = -\frac{1}{T_1} \tag{6.8}$$

The initial condition for the output of amplifier 1 is a voltage $-1/T_1$; this is indicated by I.C. $= -1/T_1$ on diagram 6.4.

We can conclude that, as the analogue units can be set up with the same system equation as the undamped remote position control system, the solution for the output position of the remote position control system can be found. The various parameters of the system equations can be altered by changing the transfer functions of the amplifiers and various potential divider adjustments.

For the system under consideration the oscillatory nature of the output is such that an output will build up without an input impulse function; this is because of small spurious voltages (noise) existing in the analogue units.

6C Example 2 of the use of an analogue computer

Simple dynamic system

The second example of the use of analogue computing units will be to find the position of a mass on a spring suspended in a liquid as shown in figure 6.5.

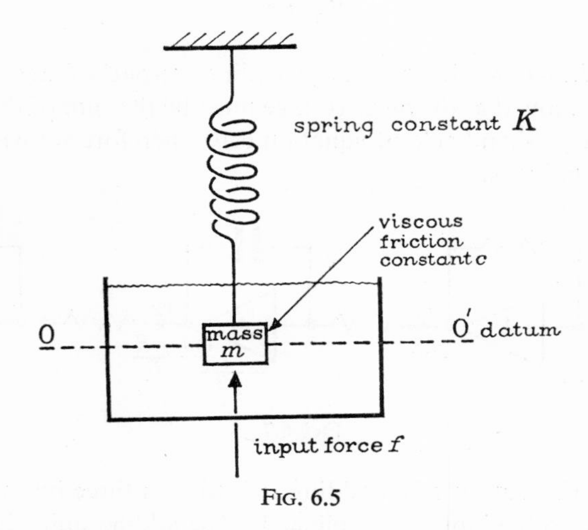

FIG. 6.5

Figure 6.5. The input to a system, often called the *forcing function*, is a force f. This force will be equal to the system mass m multiplied by the system acceleration p^2X, where X is the displacement from the datum OO′. The force f will be opposed by a viscous friction force pcX, proportional to the system velocity pX, and the spring tension KX, proportional to the displacement, where K is the spring constant. The system equation will be

$$mp^2X = f - pcX - KX \tag{6.9}$$

mass × acceleration = forcing function − viscous friction force − spring tension

If we divide equation 6.9 by the mass m it will be in a form suitable for setting up with analogue computer elements. We wish to find the displacement X at any time t. Let a voltage E be numerically equal to X, thus equation 6.9 becomes

$$p^2E = \frac{f}{m} - \frac{c}{m}pE - \frac{K}{m}E \tag{6.10}$$

As we are unable to use a practical differentiating unit we must work 'backwards' and let the input to a series of integrators be the voltage p^2E.

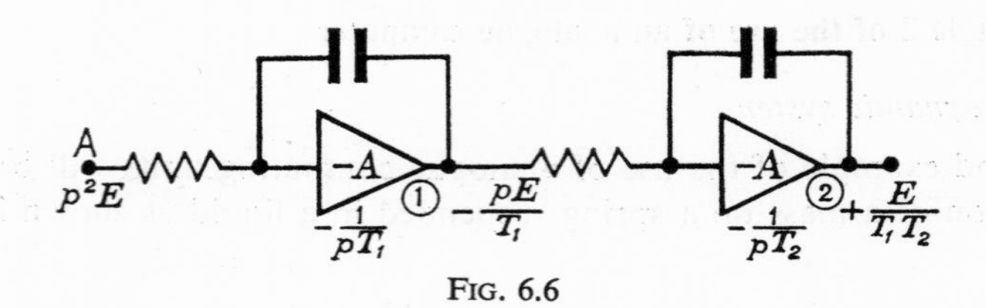

FIG. 6.6

Figure 6.6 shows two integrators providing output voltages of the form pE and E. We know that the *input* voltage must be the sum of the three terms shown on the right-hand side of equation 6.10, therefore we will connect an adding unit to point A.

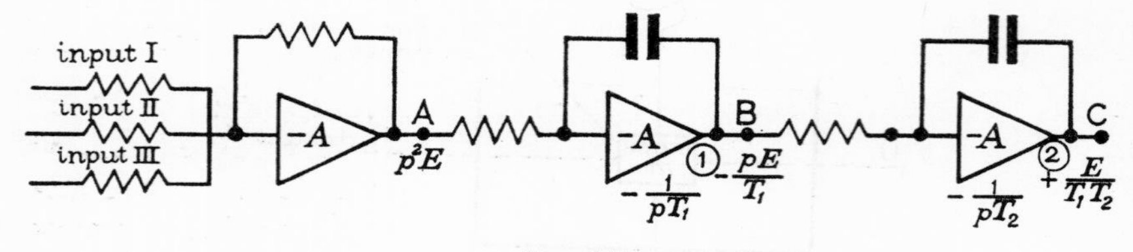

FIG. 6.7

Figure 6.7. The output of an adding unit to sum three input voltages has been connected to the input of amplifier 1. The adding unit's input voltages will be of the form $-f/m$, $+(pcE)/m$ and $+KE/m$. The individual inputs will give the appropriate transfer function to obtain the constant coefficients. The adding unit will invert the inputs to give equation 6.10.

The input function f will be obtained from an outside source and applied to input I. A voltage of the form $+pE$ is provided via an inverter from point B in the circuit in figure 6.7. A voltage of the form $+E$ is obtained from point C. Let us assume for the moment that the three individual transfer functions of the adder (amplifier 3) have been found. The complete circuit is shown in figure 6.8.

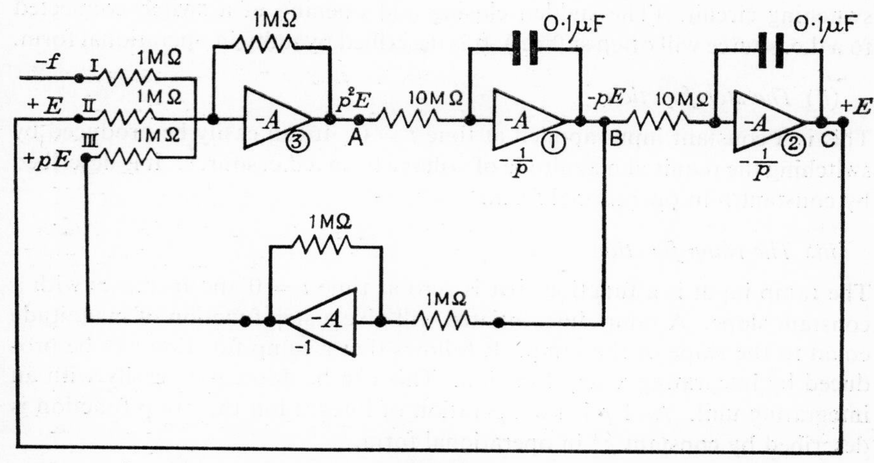

FIG. 6.8

Figure 6.8. The integrating amplifiers' time constants have been set so that $T_1 = 1$ and $T_2 = 1$. Their transfer functions become $-1/p$. The inverter has a transfer function of -1. To describe the circuit completely we must find the three transfer functions of the adder, amplifier 3. Let these transfer functions be Φ_1, Φ_2 and Φ_3 for the three inputs.

$$\Phi_1 = \frac{f/m}{-f} = -\frac{1}{m}$$

$$\Phi_2 = -\frac{K/m}{E}E = -\frac{K}{m}$$

$$\Phi_3 = -\frac{cp/m}{pE}E = -\frac{c}{m}$$

The individual transfer functions are given by the ratio of the feed-back resistor to the resistor of the input being considered. Suitable component values are shown in figure 6.8.

6D Forcing functions

The forcing functions or system inputs most commonly used are the impulse function, the step function, the ramp function and the sine function.

(*i*) *The impulse function*

This is an input occurring at time $t = 0$ only. It is dealt with in more detail in chapter 3. An approximate impulse can be produced with an electrical switching circuit. (The sudden closing and opening of a switch connected to a d.c. source will often suffice.) It is described by unity in operational form.

(ii) The step function

This is a constant input applied at time $t = 0$. It can easily be produced by switching the required magnitude of voltage from a d.c. source. It is described by constant/p in operational form.

(iii) The ramp function

The ramp input is a function that is zero at time $t = 0$ and increases with a constant slope. A ramp differentiated will give a step function of magnitude equal to the slope of the ramp. It follows that a ramp function can be produced by integrating a step function. This can be done quite easily with an integrating unit. As $1/p$ is the operation of integration the ramp function is described by constant/p^2 in operational form.

(iv) The sine function

The sine input is a function which varies sinusoidally with time t; it need not necessarily be zero at time $t = 0$. The analogue circuit shown for setting up the undamped remote position control system will produce a sine function. It can be described by $\omega/(p^2+\omega^2)$ in operational form, ω being the frequency in radians per second.

6E Scaling

We have seen how to set up system equations on the analogue computer, but we must take into account the scaling of the system equation. There are two scales to be taken into consideration—the *amplitude* and the *time* scales. The system equation and the computer system equation must obviously give the same solution of behaviour, but we can scale the problem. Amplitude scaling will ensure comparable inputs to each analogue unit without amplifier overloading. Time scaling enables us to observe the behaviour of a system in far less than the usual time, i.e. the problem is solved faster. Scaling can be complicated, so to make it easier to follow we will take the scaling of a remote position control system equation as an example.

Let the system equation be

$$5p^2\Theta_o+20p\Theta_o+80\Theta_o = \Theta_i \tag{6.11}$$

where Θ_i and Θ_o are in radians. We must choose a value for the forcing function (input position) Θ_i. Let Θ_i be a step function of magnitude 1 radian. We will consider the scaling of this system equation in two parts.

(i) Amplitude scaling

Let the system variable be Θ_o radians and the computer system variable be E_o, a voltage. It is obvious that Θ_o must be proportional to E_o.

Let

$$E_o = a\Theta_o$$

$$\frac{E_o}{\Theta_o} = a \tag{6.12}$$

where a is the amplitude scale factor in volts per radian. Let us assume that the amplifiers of the analogue computer we are using overload with a positive voltage greater than $+11$ volts and a negative voltage greater than -11 volts. Overloading is undesirable, therefore we must specify a maximum value of amplitude scale factor.

$$a_{\text{max}} = \frac{\text{maximum amplifier voltage}}{\text{maximum system variable}} \tag{6.13}$$

Let us also draw the connections between analogue units for the solution of an equation of the form of equation 6.11. We will label the output and input with their operator forms but ignore the coefficients; they will be adjusted at the adder (amplifier 3).

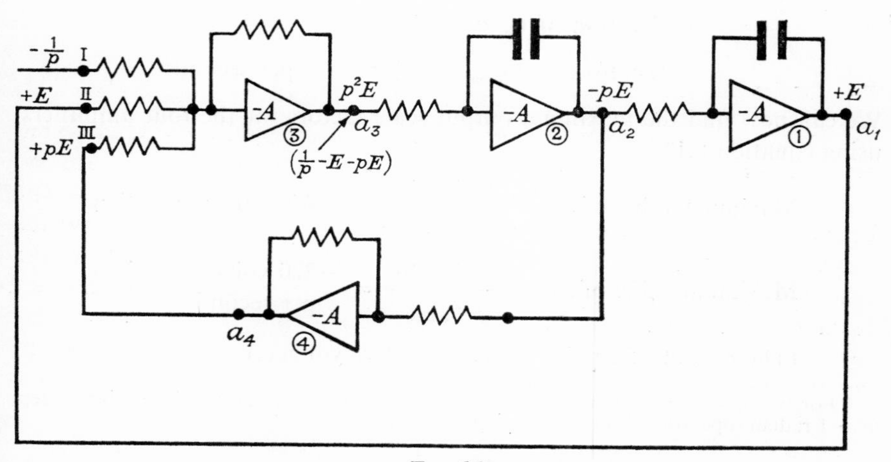

FIG. 6.9

Figure 6.9. (The circuit is of the form of figure 6.8.) The letters a_1, a_2, a_3 and a_4 stand for the amplitude scale factors for the output of each amplifier.

Equation 6.13 suggests that we must know the maxima of the system variables before being able to scale the computer system equation. We must find possible maxima for the output position Θ_o radians, the output velocity $p\Theta_o$ radians per second, and the output acceleration $p^2\Theta_o$ radians per second. We know that a remote position control system of the kind described by equation 6.11 would give a maximum output of twice the steady-state output.† (The solution for the undamped system, giving the maximum possible output, is shown in the footnote.†) The value of the input position θ_i has been taken as a step function of magnitude 1 radian. The steady-state output is found by making the operator p zero. This gives the output θ_o of $\frac{1}{80}$ radian. Therefore

$$\text{maximum position } (\theta_{omax}) = 2 \times \tfrac{1}{80}$$

$$\Theta_{omax} = \tfrac{1}{40} \text{ radian}$$

The solution, as a function of time, of equation 6.11 when undamped is

$$\theta_o = \tfrac{1}{80}(1-\cos 4t)\dagger \qquad (6.14)$$

To find the maximum output velocity, $p\Theta_{omax}$, we differentiate equation 6.14, and to find the maximum output acceleration, $p^2\Theta_{omax}$, we differentiate equation 6.14 twice. Therefore,

$$\text{velocity, } d\theta_o/dt = \tfrac{4}{80}\sin 4t$$

$$\text{therefore, } p\Theta_{omax} = \tfrac{4}{80} \text{ radians per second}$$

$$\text{acceleration, } d^2\theta_o/dt^2 = \tfrac{16}{80}\cos 4t$$

$$\text{therefore, } p^2\Theta_{omax} = \tfrac{16}{80} \text{ radians per second}^2$$

We can now find the maximum output scale factors for the four amplifiers using equation 6.13.

$$\text{Maximum value of } a_1 = \frac{+11 \text{ volts}}{\Theta_{omax}} = 440 \text{ volts per radian}$$

$$\text{Maximum value of } a_2 = \frac{-11 \text{ volts}}{p\Theta_{omax}} = -220 \text{ volts per radian per second}$$

(The output of amplifier 2 is a negative voltage.)

† For the maximum output the damping term ($p\Theta_o$) will be zero. With an input step $\theta_i = 1$ radian (operational form of the input $= 1/p$)

$$(5p^2+80)\,\Theta_o = \frac{1}{p}$$

$$\Theta_o = \frac{1}{5}\left(\frac{1}{p}\cdot\frac{1}{p^2+16}\right)$$

$$\Theta_o = \frac{1}{5}\left(\frac{1/16}{p} - \frac{p1/16}{p^2+16}\right)$$

as a function of time t, $\theta_o = 1/80\,(1-\cos 4t)$ radians

$$\text{Maximum value of } a_3 = +\frac{11 \text{ volts}}{p^2\Theta_{omax}} = \begin{array}{l}55 \text{ volts per radian}\\ \text{per second}^2\end{array}$$

Maximum value of a_4 = maximum value of a_2 say.

This means that amplifier 4 will have a unity gain. It is convenient to have the scale factors in round figures.

Let $a_1 =$ 400 volts per radian

$a_2 = -200$ volts per radian per second

$a_3 =$ 50 volts per radian per second^2

$a_4 = -a_2 = 200$ volts per radian per second

To be able to set up the analogue circuit we must know the transfer function of each amplifier. We have already specified the transfer function of amplifier 4 as being unity (-1). The transfer functions of amplifiers 1 and 2 will be $-1/pT_1$ and $-1/pT_2$ respectively. Amplifier 1 integrates *velocity* to give *position*, i.e. in the actual system as a hypothetical unit its input would be $p\Theta_o$ and its output Θ_o (coefficients are adjusted at the adding unit). The analogue system elements must have the same transfer function as the hypothetical system unit, but with consideration of the amplitude scale factor. Using equation 6.12,

$$\text{input to amplifier 1 } (\propto -pE) = a_2 p\Theta_o \tag{6.15}$$

$$\text{output of amplifier 1 } (\propto E) = a_1\Theta_o \tag{6.16}$$

The ratio of the output to the input of amplifier 1, which is its transfer function, will be

$$-\frac{1}{pT_1} = \frac{a_1\Theta_o}{a_2 p\Theta_o}$$

therefore

$$T_1 = -\frac{a_2}{a_1} = \tfrac{1}{2} \text{ second}$$

Similarly, input to amplifier 2, $(\propto p^2E) = a_3 p^2\Theta_o$

output of amplifier 2, $(\propto -pE) = a_2 p\Theta_o$

The ratio of the output to the input of amplifier 2, which is its transfer function, will be

$$-\frac{1}{pT_2} = \frac{a_2 p\Theta_o}{a_3 p^2\Theta_o}$$

therefore

$$T_2 = -\frac{a_3}{a_2} = \tfrac{1}{4} \text{ second}$$

We now have to find the three transfer functions, one for each input of the adder, amplifier 3, to give us the complete information to set up the system equation on the analogue computer. Rearranging the system equation 6.11 in a form in which it may be presented to the adding unit, we have

$$p^2\Theta_o = 0{\cdot}2\Theta_i - 4p\Theta_o - 16\Theta_o \tag{6.17}$$

Equation 6.17 gives us the coefficients 0·2, 4 and 16 relating the output and the inputs of a unity scaled adding unit. However, we must take into account the amplitude scale factors. Consider equation 6.12 in the following form,

adder output = output scale factor × system *output* coefficient
adder input = input scale factor × system *input* coefficient

The individual transfer function, Φ, will be given by the ratio of the adder output to the input,

$$\Phi = \frac{\text{output scale factor}}{\text{input scale factor}} \times \underbrace{\frac{\text{system 'output'}}{\text{system 'input'}}}$$

$$\Phi = \frac{\text{output scale factor}}{\text{input scale factor}} \times \text{system coefficient} \tag{6.18}$$

The system input coefficient is unity (the coefficient of $p^2\Theta_o$). The input position θ_i has been taken as a step function of magnitude 1 radian. Let this be a step function of −5 volts in the analogue computer. This will give another amplitude scale factor, a_5 say, of 5 volts per radian. The transfer function of input (I) is, from equation 6.18

$$\Phi_1 = \frac{a_3}{a_5} \times \frac{0{\cdot}2}{1} = \frac{50}{-5} \times 0{\cdot}2$$

$$\Phi_1 = -2$$

Similarly, for input (II) the transfer function is

$$\Phi_2 = \frac{a_3}{a_1} \times -16 = \frac{50}{400} \times -16$$

$$\Phi_2 = -2$$

Similarly, for input (III) the transfer function is

$$\Phi_3 = \frac{a_3}{a_4} \times -4 = \frac{50}{200} \times -4$$

$$\Phi_3 = -1$$

We are now able to draw the analogue computer circuit, including practical component values, with suitable amplitude scaling.

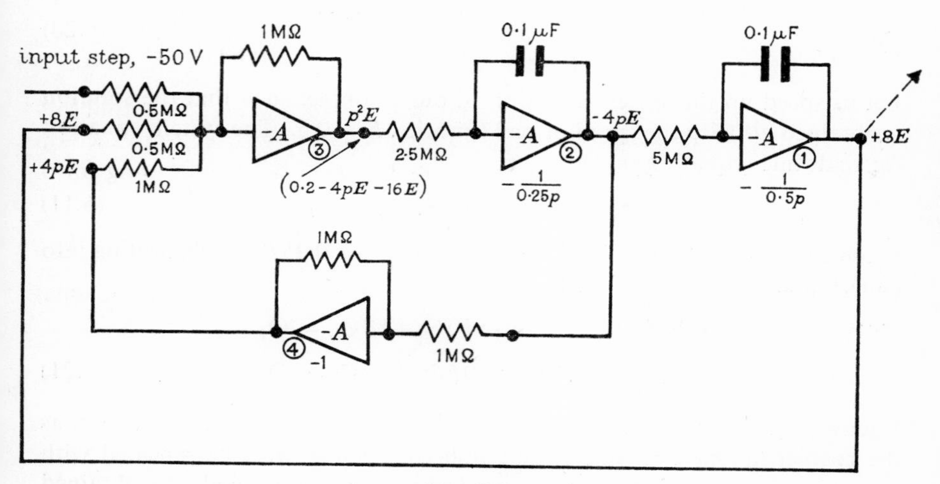

FIG. 6.10

Figure 6.10. This is a practical circuit for setting up the system equation 6.11 of a damped remote position control system. The solution, with output position θ_o suitably scaled, is obtained from the output of amplifier 1.

(*ii*) *Time scaling*

We shall now consider the same system equation as before, but will time-scale it. If a system equation is set up and not time-scaled, the solution is said to be found in *real time.*

Let t be the symbol for real time and p be the operator that operates in real time.

Let T be the symbol for the *scaled time* and P be the operator that operates in scaled time.

Let b be the time scale factor.

$$T = bt$$

therefore

$$b = \frac{T}{t}$$

$$P \equiv \frac{d}{dT} \quad \text{and} \quad p \equiv \frac{d}{dt}$$

therefore
$$b = \frac{p}{P} \tag{6.19}$$

and
$$b^n = \frac{p^n}{P^n} \tag{6.20}$$

Let us speed up the system equation of our example, such that the solution is obtained at twice the speed of real time, i.e. $b = 0{\cdot}5$. The system equation for real time (equation 6.11) is

$$5p^2\Theta_o + 20p\Theta_o + 80\,\Theta_o = \Theta_i \tag{6.11}$$

From equations 6.19 and 6.20, $p = 0{\cdot}5P$ and $p^2 = 0{\cdot}25P^2$. Substituting into equation 6.11, we obtain

$$5(0{\cdot}25P^2)\Theta_o + 20(0{\cdot}5P)\Theta_o + 80\,\Theta_o = \Theta_i$$

$$1{\cdot}25P^2\Theta_o + 10P\Theta_o + 80\,\Theta_o = \Theta_i \tag{6.21}$$

Equation 6.21 is now set up on the analogue computer. We can see that, as the coefficients have changed, the amplitude scaling must be repeated with the new system equation. A time-scaled system equation can also be obtained by changing every time constant by the time scale factor *b*.

6F Comments and conclusions

A linear differential system equation can be set up on an analogue computer as follows:

(*i*) The equation is rearranged so that the term containing the highest order of the operator *p* is alone on the left-hand side.

(*ii*) The left-hand side of the equation is called the *input*. This is applied to a series of integrators to give the required proportional operational terms shown on the right-hand side of the system equation.

(*iii*) The terms are applied with the consideration of sign, via an adder to the input.

(*iv*) Initial conditions are set up on the integrators.

(*v*) The system equation is time-scaled if this is necessary.

(*vi*) The system equation is amplitude-scaled to give the correct time constants and transfer functions of the units.

Often the problem of setting up system equations may seem difficult because of the size of one coefficient with respect to another, e.g. in the equation $2p^2\Theta_o + 100p\Theta_o + 10\,\Theta_o = \Theta_i$ the second-order term's coefficient is small (2) compared with the first-order term's coefficient (100). Time scaling can often equalize the coefficients.

The techniques involved in analogue computing are complicated and we have only considered simple practical techniques, but they should be useful as a basis for experimental work and further study.

Non-linear elements for analogue computing are discussed in the appendix.

6G Examples

1. Show analytically: (*a*) how a simple *RC* circuit may be used to obtain an output voltage which is approximately the time-integral of an input voltage; and (*b*) how the performance of this integrator may be improved by using a suitable high-gain d.c. amplifier.

Hence give in a diagram, with brief comments, the essential elements and connections for an electronic analogue computing system that could be used to solve an equation of the form

$$a\frac{d^2x}{dt^2}+b\frac{dx}{dt}+cx = 0$$

where a, b and c are constants and x is a variable.

(Part 3. I.E.E. 1962)

2. Describe, with diagrams and comments, how a simple dynamic system can be simulated on an analogue computer, and discuss the advantages of the simulation with references to your example.

3. A remote position control system is described by the closed-loop transfer function Φ, where

$$\Phi = \frac{\omega_n^2}{p^2+2\zeta\omega_n p+\omega_n^2}$$

Show how this transfer function can be set up on an analogue computer. Indicate initial conditions clearly on an analogue computing circuit.

4. A control system of the type in question 3 has a damping ratio $\zeta = 0{\cdot}5$ and an undamped natural frequency of 10 radians per second. Draw a diagram of an analogue computing circuit using two integrating amplifiers, both of transfer function $1/(pT_1)$ and one adding amplifier, each input having a unity transfer function. Find the initial state of the integrating amplifiers.

$(0;\ -1/T_1)$

5. Describe the steps involved in setting up the third-order linear differential equation

$$y = 3\frac{d^3x}{dt^3}+2\frac{d^2x}{dt^2}+\frac{dx}{dt}+4x$$

on an analogue computer.

6. Show how the following functions can be produced, in the form of time-varying voltages, by an analogue computer.

(*i*) $f(t) = A$ where A is constant

(*ii*) $f(t) = 3t$

(*iii*) $f(x) = 5x^3+x$

(*iv*) $f(t) = \cos(\omega t+\phi)$ where ω and ϕ are constant

(*v*) $f(x) = (1+x)\,e^{-x}$

(*vi*) $f(x) = 10\sin(5x+60°)$

(Initial conditions are defined by the function.)

7. Discuss the reasons for scaling analogue computers and describe the scaling of a simple problem.

8. An analogue computer's amplifiers overload with voltages greater than 20 volts and less than −20 volts. Set up the following transfer functions on an analogue computer, scaling each one.

(*i*) $$\Phi = \frac{10}{p^2+16}$$

(*ii*) $$\Phi = \frac{9}{p^2+6p+9}$$

(*iii*) $$\Phi = \frac{16}{p^2+16p+16}$$

(*iv*) $$\Phi = \frac{4}{p(p+2)(p+3)}$$

CHAPTER 7

The Root-Locus Pattern

7.1 The s-plane and the $|\overline{\Phi}|$ - surface

Students who have not included chapter 3 in their reading should note that for control systems we shall consider operator p and the complex variable s as being identical. s is the complex variable $\alpha+j\omega$, where α and ω are quantities having the dimensions of 1/time and j is the complex operator $\sqrt{-1}$. All transfer functions in the text will now be written in terms of s instead of p. A bar above a symbol will indicate that it is a function of s.

We can express a general control system transfer function $\overline{\Phi}$ as follows

$$\overline{\Phi} = \frac{K(s^n+a_{n-1}s^{n-1}+a_{n-2}s^{n-2}+\ldots+a_0)}{(s^m+b_{m-1}s^{m-1}+b_{m-2}s^{m-2}+\ldots+b_0)}$$

where K is a constant. $a_0, a_1, a_2, \ldots a_{n-1}$ and $b_0, b_1, b_2 \ldots b_{m-1}$ are constant coefficients and m is greater than n. The general transfer function $\overline{\Phi}$ can also be expressed as

$$\overline{\Phi} = \frac{K(s-r_1)(s-r_2)(s-r_3)\ldots(s-r_n)}{(s-q_1)(s-q_2)(s-q_3)\ldots(s-q_m)} \tag{7.1}$$

The roots $r_1, r_2, r_3, \ldots r_n$ and $q_1, q_2, q_3, \ldots q_m$ may be complex, real or zero. We shall now define two singularities of $\overline{\Phi}$.

(*i*) A *pole* of $\overline{\Phi}$ is a value of s which makes the denominator zero, i.e. $\overline{\Phi} = \infty$. An example of a pole is $s = q_1$.

(*ii*) A *zero* of $\overline{\Phi}$ is a value of s which makes the numerator zero, i.e. $\overline{\Phi} = 0$. An example of a zero is $s = r_1$.

Let us consider a specific system transfer function to illustrate poles and zeros.

Let
$$\overline{\Phi} = \frac{K(s+3)}{s(s+2+j4)(s+2-j4)} \tag{7.2}$$

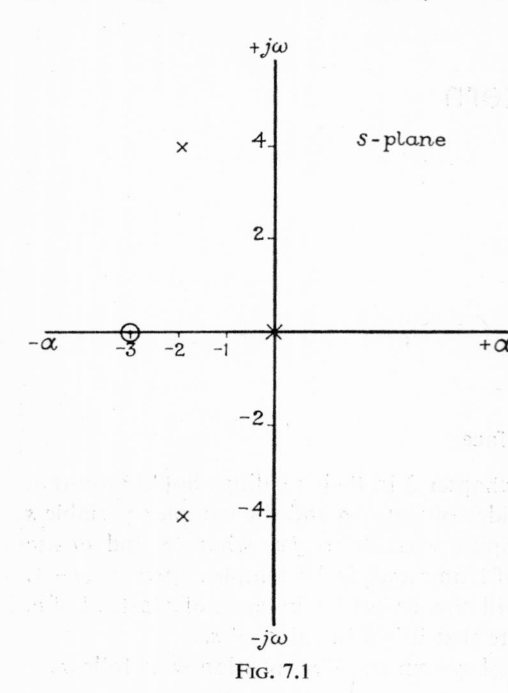

FIG. 7.1

From the definitions of poles and zeros equation 7.2 has three poles at $s=0$, $s=-2-j4$ and $s=-2+j4$.

Poles and zeros can be plotted on an *s-plane*. An *s*-plane consists of a real axis α and an imaginary axis ω.

Figure 7.1. This figure shows an *s*-plane on which the poles and zeros of equation 7.2 have been plotted. A pole is represented by a small cross and a zero by a small circle.

A general transfer function $\overline{\Phi}$ has a different particular value for every value of the complex variable *s*. Let a particular value of *s* be $\alpha_1+j\omega_1$ and the corresponding value of $\overline{\Phi}$ be A_1+jB_1, where A_1 and B_1 are constants.

$$[\overline{\Phi}]_{s=\alpha_1+j\omega_1} = A_1+jB_1$$

Expressed in polar form

$$[\overline{\Phi}]_{s=\alpha_1+j\omega_1} = \sqrt{(A_1^2+B_1^2)}\angle\theta$$

$$[\overline{\Phi}]_{s=\alpha_1+j\omega_1} = |\overline{\Phi}|_{s=\alpha_1+j\omega_1}\angle\theta \tag{7.3}$$

Equation 7.3 shows that we have three quantities to consider α, ω and $\overline{\Phi}$. We shall just consider α, ω and the modulus of the transfer function $|\overline{\Phi}|$; phase θ will be dealt with later. A third axis is required. This axis is drawn mutually perpendicular to the α and ω axes.

Figure 7.2. The *three-dimensional* axes are shown in this figure.

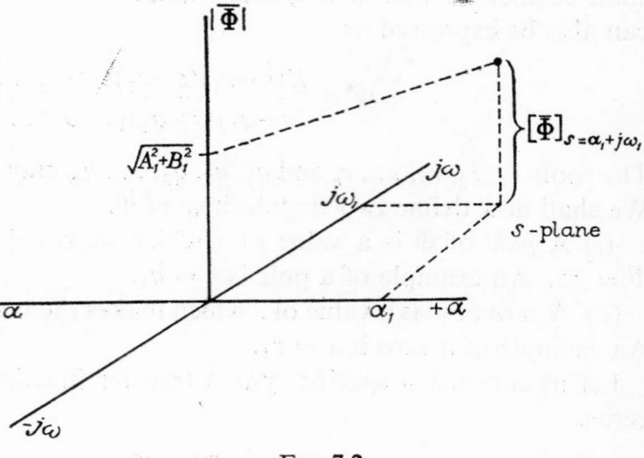

FIG. 7.2

The modulus of $\overline{\Phi}$ is shown for the value of $s=\alpha_1+j\omega_1$. A $|\overline{\Phi}|$-surface can be plotted on the three-dimensional axes. To clarify this and to see the effect of poles and zeros on the $|\overline{\Phi}|$-surface we shall consider an example.

Let a system transfer function be

$$\overline{\Phi}=\frac{2}{s+1} \tag{7.4}$$

By definition the transfer function has a pole at $s=-1$ and no finite zeros. For simplicity we shall first draw the s-plane *in two dimensions* and consider a plot of the $|\overline{\Phi}|$-surface afterwards.

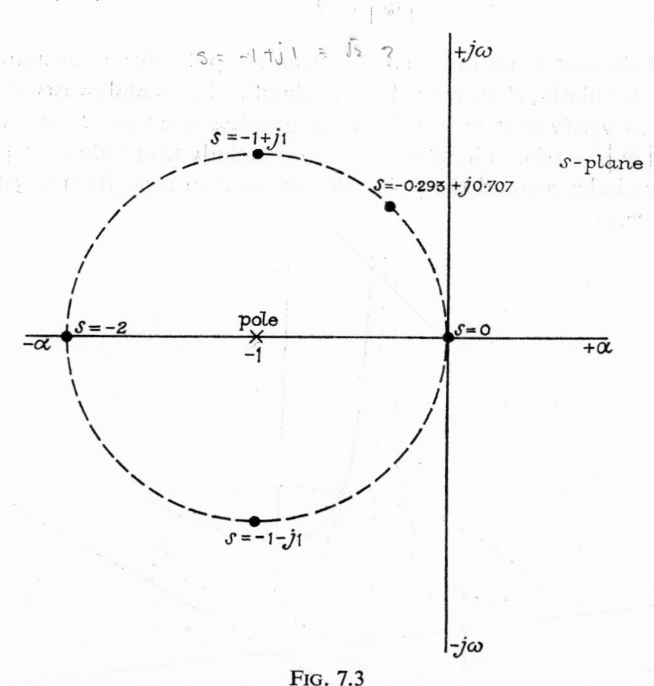

FIG. 7.3

Figure 7.3. This figure shows the s-plane with the pole $s=-1$ marked on it. A circle of unity radius and centre at the pole has also been drawn on the s-plane. Consider values of the modulus of the transfer function $|\overline{\Phi}|$ for values of s on the circumference of this circle.

$s=0$:

$$\overline{\Phi}=2$$

$$|\overline{\Phi}|=2$$

$s=-0{\cdot}293+j0{\cdot}707$:

$$\overline{\Phi}=\frac{2}{0{\cdot}707+j0{\cdot}707}$$

$$|\overline{\Phi}|=2$$

$s = -1+j1$:

$$\bar{\Phi} = \frac{2}{j1}$$

$$|\bar{\Phi}| = 2$$

$s = -2$:

$$\bar{\Phi} = \frac{2}{-2+1}$$

$$|\bar{\Phi}| = 2$$

Thus the values of s at a unit distance from the pole give a constant value of $|\bar{\Phi}| = 2$. Similarly, if we consider a circle of radius 2 and centre at the pole, it is easy to verify that we shall obtain another constant value of $|\bar{\Phi}|$ (in this case $|\bar{\Phi}| = 1{\cdot}0$). Therefore we can conclude that values of $|\bar{\Phi}|$ form concentric circles around the pole. We are now in a position to *construct* a $|\bar{\Phi}|$-surface.

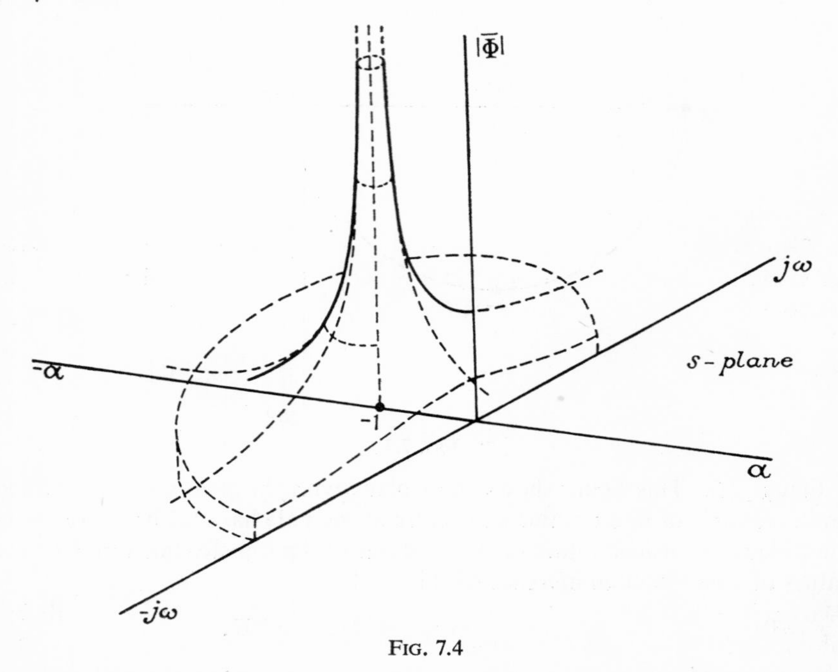

FIG. 7.4

Figure 7.4. The $|\bar{\Phi}|$-surface for the transfer function $2/(s+1)$ is shown in this figure. The surface goes to infinity at the pole and it becomes zero when

s is infinite. There is a zero for every pole. Therefore, the transfer function $2/(s+1)$ has one non-finite zero.

The phase of a value of $\overline{\Phi}$ can be indicated on the $|\overline{\Phi}|$-surface. Consider a line drawn parallel to the ω-axis and through the pole on the s-plane.

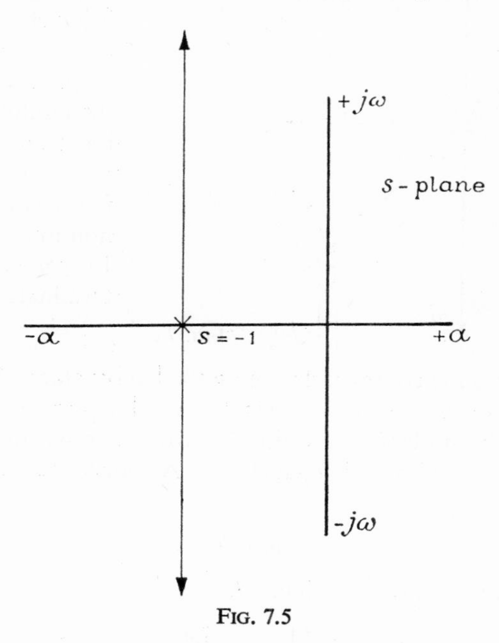

FIG. 7.5

Figure 7.5. We shall consider values of the phase θ, for different values of s, along a line drawn parallel to the ω-axis and through the pole on the s-plane.

$s = -1+j0{\cdot}001$: $$\overline{\Phi} = |\overline{\Phi}| \angle\theta = \frac{2}{j0{\cdot}001}$$

therefore $\theta = -90°$

$s = -1+j1000$: $$\overline{\Phi} = |\overline{\Phi}| \angle\theta = \frac{2}{j1000}$$

therefore $\theta = -90°$

We can see that values of θ going in a positive direction away from the pole along the line are constant at $-90°$. Similarly, values of θ going in a negative direction away from the pole along the line are constant at $+90°$. We can quite easily verify that radial lines from the pole are at a constant phase and thus indicate the phase of any value of $\overline{\Phi}$.

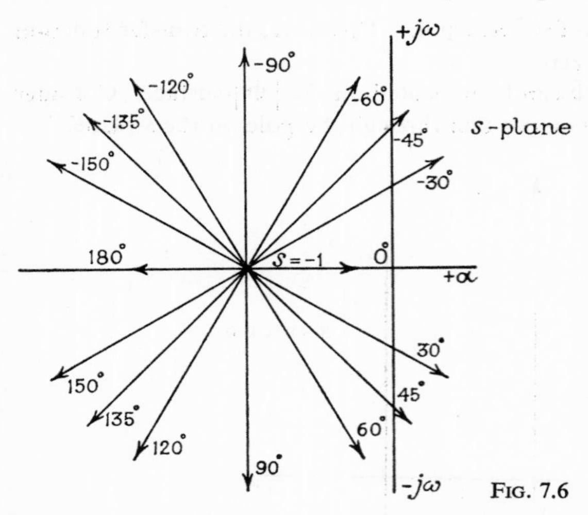

FIG. 7.6

Figure 7.6 shows the phase lines. We can draw the circles and the phase lines on the $|\overline{\Phi}|$-surface in much the same way as contours are drawn on a map. The contours specify the modulus and phase of the transfer function on the $|\overline{\Phi}|$-surface. The phase contours go to the non-finite zero at infinity. In general, phase lines terminate on a zero.

Figure 7.7 shows the contours drawn on the $|\overline{\Phi}|$-surface. We have chosen a very simple transfer function to illustrate the $|\overline{\Phi}|$-surface, the shape and contours having been obtained intuitively. More complicated transfer functions have more complicated $|\overline{\Phi}|$-surfaces. A transfer function of the form

$$\frac{1}{(s+a-jb)(s+a+jb)} \tag{7.5}$$

has a $|\overline{\Phi}|$-surface as shown in Figure 7.8.

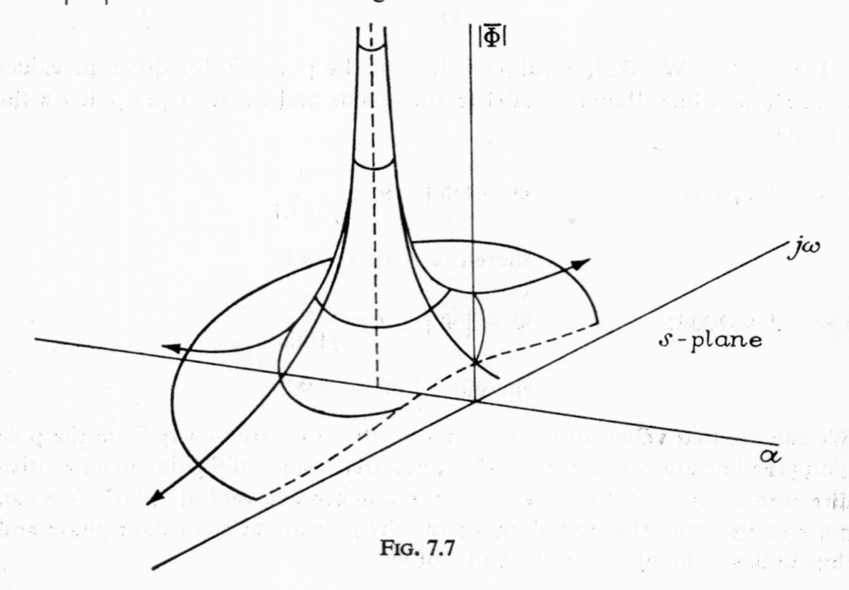

FIG. 7.7

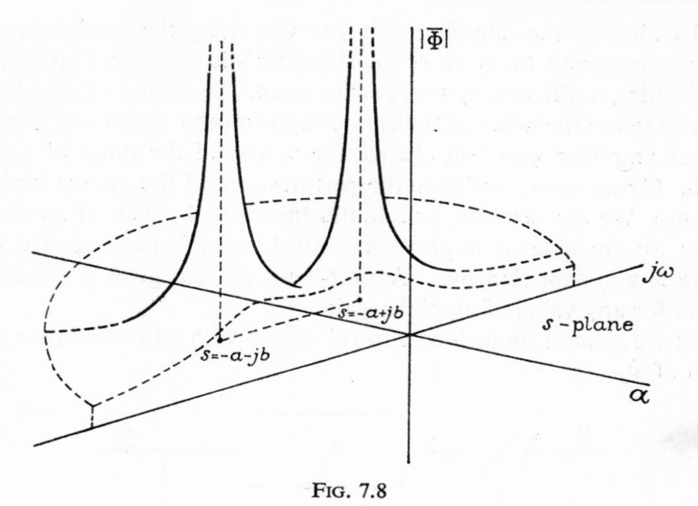

FIG. 7.8

Figure 7.8. There are two poles at $s = -a+jb$ and $s = -a-jb$. (These roots are called conjugate complex roots. As transfer functions always have real coefficients, complex parts always appear as a conjugate pair.) There are no finite zeros. The $|\bar{\Phi}|$-contours are circles near the poles, the phase contours start at a pole and terminate at a zero (non-finite in this case, a finite zero is like a 'pit' on the $|\bar{\Phi}|$-surface).

The physical significance of the $|\bar{\Phi}|$-surface is not relevant to a discussion of the root-locus pattern, but it is hoped that its description will help students in their appreciation of what is to follow. The root-locus pattern makes use of the 180° phase contour.

Points to bear in mind before we begin the root-locus method are:

(*i*) Phase contours on the *s*-plane start at poles and terminate at zeros.
(*ii*) $|\bar{\Phi}|$ and phase contours always cut at right angles.

7.2 The root-locus

The transfer function of the second-order remote position control system described in chapter 4 is given by equation 4.26

$$\frac{\bar{\theta}_o}{\bar{\theta}_i} = \frac{\omega_n^2}{s^2+2\zeta\omega_n s+\omega_n^2}$$

The factor ω_n^2 is $(mK_1K_2)^2$ where m is the gain of an electronic amplifier, K_1 is the transfer function of a potentiometer and K_2 is the transfer function of a servo-motor. (ω_n represents the undamped natural frequency of the system.) If K_1 and K_2 are constants, the coefficients of the transfer function can be varied by altering the amplifier gain m. Changing the coefficients of the denominator changes the roots of the characteristic equation $s^2+2\zeta\omega_n s+\omega_n^2$, thus indicating a different system performance. The roots of the characteristic equation are the poles of the closed-loop transfer function. Therefore a change of amplifier gain will change the values of the poles of a transfer function. (These values indicate the performance of the system under consideration.) We can see that by an adjustment of the gain of an electronic amplifier we are able to improve a control system's performance without changing any system elements. A root-locus pattern gives a system's performance for any value of amplifier gain.

Consider a general single-loop control system with a forward path transfer function of $\overline{\Phi}_a$.

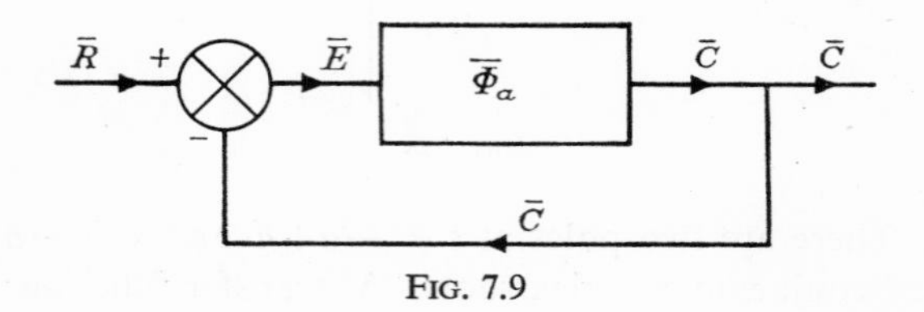

FIG. 7.9

Figure 7.9. The system is shown in this figure. Its closed-loop transfer function is

$$\overline{\Phi} = \frac{\overline{C}}{\overline{R}} = \frac{\overline{\Phi}_a}{1+\overline{\Phi}_a} \tag{7.6}$$

Let $\overline{\Phi}_a = K\overline{\Phi}_1$ where K is a gain constant. [The gain constant for the transfer function (equation 4.26) is $(mK_1K_2)^2$.] In general the gain constant is the scalar part of a transfer function as shown in equation 7.1. The components contributing to the gain constant are contained in the block of the forward path. Equation 7.6 can be rewritten

$$\overline{\Phi} = \frac{K\overline{\Phi}_1}{1+K\overline{\Phi}_1} \tag{7.7}$$

$\overline{\Phi}$ is the ratio of two polynomials in s. The denominator of the ratio characterizes the system. The closed-loop poles are given by

$$1+K\overline{\Phi}_1 = 0 \tag{7.8}$$

We can see that the closed-loop poles characterize the system, thus equation 7.8 is the characteristic equation of the system. Rearranging equation 7.8 we have

$$\overline{\Phi}_1 = -\frac{1}{K} \tag{7.9}$$

$\overline{\Phi}_1$ is a function of the complex variable s and equation 7.9 can be expressed in the polar form,

$$\overline{\Phi}_1 = \left|\frac{1}{K}\right| \angle 180^\circ \tag{7.10}$$

Equation 7.10 shows that the modulus of $\overline{\Phi}_1$ is $|1/K|$ and its phase is 180° (or $180^\circ + N360^\circ$, where $N = 0, 1, 2, 3$, etc.). s, for this system, can have any value as long as equation 7.10 is satisfied. $K\overline{\Phi}_1$ is the open-loop transfer function and is the ratio of two polynomials in s. Equation 7.10 indicates that, if we draw the pole-zero pattern for the open-loop transfer function $\overline{\Phi}_1$ and the 180° phase contour, we have a locus defining the closed-loop system. The plot is called a *root-locus pattern* because equation 7.10 is factorized to give its roots in order to obtain the open-loop pole-zero pattern. As the root-locus pattern defines a system in the same way that the characteristic equation defines it, we can use a root-locus pattern to provide information about a system. The first thing that we should note, from equation 7.10, is that a change in the gain constant gives a different value of $\overline{\Phi}_1$ on the root-locus. However, before understanding the significance of the root-locus pattern, we must be able to plot root-loci. The previous section on the $|\overline{\Phi}|$-surface indicates that the plotting of root-loci for second-order systems, and above, is no easy matter, but it is possible to develop a set of rules for the quick construction of root-loci. The rules follow from a series of properties of root-loci given in note form below. Where necessary each property is proven.

(*a*) The general open-loop transfer function (equation 7.1) is of the form

$$K\overline{\Phi}_1 = \frac{K(s-r_1)(s-r_2)(s-r_3)\ldots(s-r_n)}{(s-q_1)(s-q_2)(s-q_3)\ldots(s-q_m)}$$

n gives the number of finite zeros and m gives the number of poles. The number of zeros and poles must be equal, thus $(m-n)$ gives the number of non-finite zeros. i.e. infinite zeros

(*b*) A phase contour is a continuous curve which starts at a pole and finishes at a zero; thus there are as many root-loci on a root-locus pattern as poles.

(*c*) To plot the 180° phase contour it is necessary to know the angle at which the contour leaves a pole and arrives at a zero. The direction of a root-locus can be found as follows. Let us consider a pole-zero pattern for an open-loop transfer function

$$\overline{\Phi}_a = \frac{K(s+r_1)}{(s+q_1)(s+q_2)(s+q_3)} = KG \tag{7.11}$$

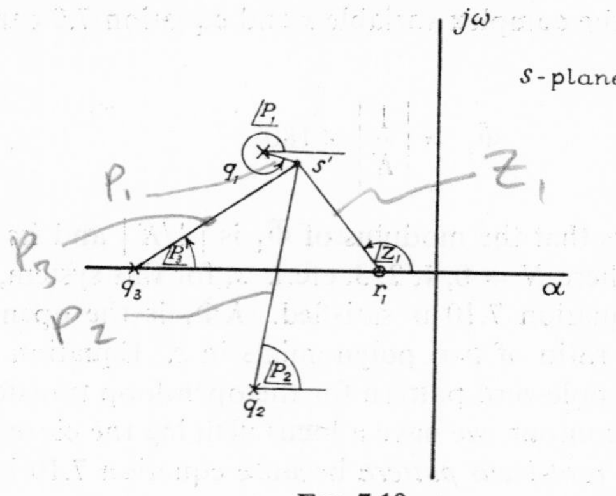

FIG. 7.10

Figure 7.10. The figure shows a possible pole-zero pattern on the s-plane. We know that a root-locus starts at a pole. Consider pole q_1 and a value of s, say s', near to q_1. The value of the open-loop transfer function at $s = s'$ is given by

$$[\Phi_a]_{s=s'} = \frac{K\mathbf{Z}_1}{\mathbf{P}_1\mathbf{P}_2\mathbf{P}_3} \tag{7.12}$$

where $\mathbf{Z}_1$ is the vector from zero r_1 to s', and $\mathbf{P}_1$, $\mathbf{P}_2$ and $\mathbf{P}_3$ are the vectors from poles q_1, q_2 and q_3 respectively, to s'. For the point s' to be on the root-locus the phase angle of $[\Phi_a]_{s=s'}$ must be 180°, i.e.

$$[|\Phi_a| \angle 180°]_{s=s'} = \frac{K|Z_1|\angle Z_1}{|P_1||P_2||P_3|(\angle P_1 + \angle P_2 + \angle P_3)}$$

where the vectors $\mathbf{Z}_1$, $\mathbf{P}_1$, $\mathbf{P}_2$ and $\mathbf{P}_3$ have phase angles $\angle Z_1$, $\angle P_1$, $\angle P_2$ and $\angle P_3$, respectively. (All angles are measured in an anti-clockwise direction from the $+\alpha$-axis.) From the above equation

$$180° = \angle Z_1 - (\angle P_1 + \angle P_2 + \angle P_3)$$

but $\angle P_1$ is the angle of departure of the root-locus from the pole q_1. Therefore, this angle of departure is

$$\angle P_1 = \angle Z_1 - (\angle P_2 + \angle P_3) - (180° + N360°) \tag{7.13}$$

We assume that the point s' is so close to the pole q_1 that the vectors $\mathbf{Z}_1$, $\mathbf{P}_2$ and $\mathbf{P}_3$ may be considered as originating from the zero r_1, pole q_2 and pole q_3 respectively to the pole q_1. In general angle of departure of a root-locus from an open-loop pole $= \Sigma \angle Z - \Sigma \angle P - (180° + N360°)$.

It follows that to find points on a root-locus the following relationship must be satisfied,

$$180° + N360° = \Sigma \angle Z - \Sigma \angle P \qquad (7.14)$$

where $\Sigma \angle Z$ is the sum of the phase angles of the vectors from the zeros to the point and $\Sigma \angle P$ is the sum of the phase angles from the poles. As before all angles are measured in an anti-clockwise direction from the $+\alpha$-axis and $N = 0, 1, 2, 3$, etc.

(*d*) The gain constant K is zero at the poles and infinity at the zeros of an open-loop transfer function. This can be seen by examination of equation 7.9,

$$K\overline{\Phi}_1 = -1$$

If K tends to zero $\overline{\Phi}_1$ must tend to infinity, i.e. s must tend to the value of the poles of $\overline{\Phi}_1$. If K tends to infinity $\overline{\Phi}_1$ must tend to zero, i.e. s must tend to the value of the zeros of $\overline{\Phi}_1$ [this is also a proof of note (*b*)].

(*e*) Some root-loci will be along the real axis.

Consider a pole-zero pattern for an open-loop transfer function.

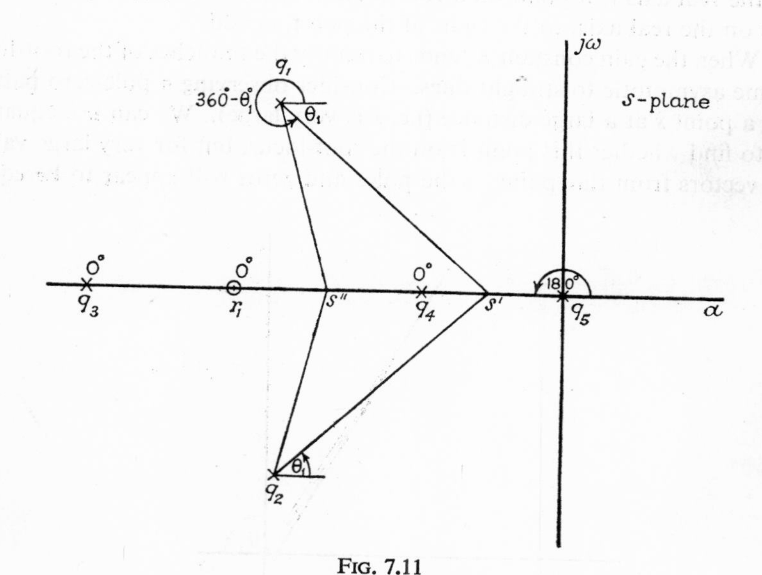

FIG. 7.11

Figure 7.11. Two points, s' and s'', on the real axis will be considered. Applying equation 7.14 to the point s', we can find the angle of departure of the root-locus from q_5. If the point s' is on a root-locus the angle of departure should be 180°. Let the angle of departure be θ'. From equation 7.14,

$$180^\circ + N360^\circ = \Sigma \text{ angles of the vectors from the zeros to } s' - \Sigma \text{ angles of the vectors from the poles to } s'$$
$$180^\circ + N360^\circ = (0^\circ) - [(360^\circ - \theta_1) + \theta_1 + 0^\circ + 0^\circ + \theta')]$$
$$540^\circ + N360^\circ = -\theta'$$
$$\text{therefore} \quad \theta' = 180^\circ$$

Therefore point s' is on a root-locus. It is quite easy to show that all points from pole q_5 to pole q_4 are also on this root-locus.

Applying equation 7.14 to the point s'', we can find the angle of departure of the root-locus from q_4. Let the angle of departure be θ''. Again from equation 7.14,

$$180^\circ + N360^\circ = (0^\circ) - [(360^\circ - \theta_2) + \theta_2 + 0^\circ + 180^\circ + \theta'']$$
$$720^\circ + N360^\circ = -\theta''$$
$$\text{therefore} \quad \theta'' = 0^\circ$$

Therefore point s'' is not on a root-locus. Using equation 7.14 we are able to verify, for a pole-zero pattern with roots on the real axis, whether a root-locus occurs on the real axis or not. A general rule can be found: a value of s on the real axis is a point on a root-locus if the total number of poles and zeros on the real axis, to the right of the point, is odd.

(*f*) When the gain constant K tends to infinity the branches of the root-locus become asymptotic to straight lines. Consider observing a pole-zero pattern from a point s at a large distance (i.e. s is very large). We can use equation 7.14 to find whether this point is on the root-locus, but for very large values of s, vectors from this point to the poles and zeros will appear to be equal.

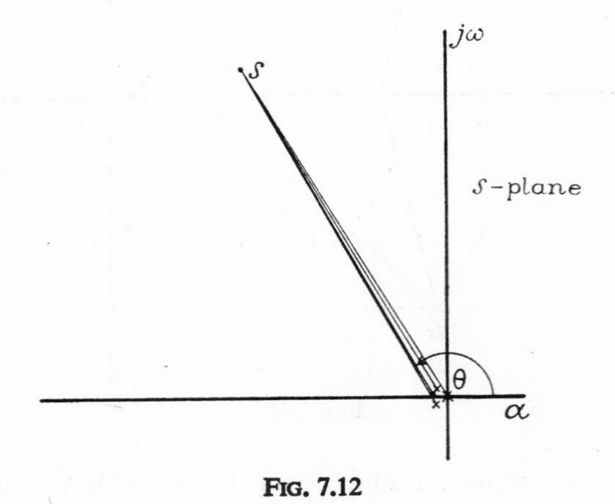

FIG. 7.12

Figure 7.12. In this figure point s is a large distance from the pole-zero pattern. The pole-zero pattern appears as a cluster of points. The vectors from the poles and zeros to point s can be assumed to be the same length and the phase angle can be considered the same for all the vectors. We can verify whether point s is on a root-locus using equation 7.14,

$$\underset{\text{(the sum of the angles of the vectors from the zeros)}}{\text{number of zeros} \times \theta} - \underset{\text{(the sum of the angles of the vectors from the poles)}}{\text{number of poles} \times \theta} = 180° + N360° \quad (7.15)$$

If n represents the number of finite zeros and m the number of poles, a point s at a large distance from the pole-zero pattern will be on a root-locus if, from equation 7.15,

$$\theta = \frac{180° + N360°}{n - m} \quad (7.16)$$

By definition, if s is at a non-finite zero, equation 7.16 will be true and angle θ can be found. We can see that, as s is brought along the root-locus from the non-finite zero towards the pole-zero pattern, the angle θ will be unchanged. This will be true until s comes close enough to the pole-zero pattern for the pattern to be distinguishable from a close cluster of points. We can conclude that the root-locus is asymptotic to a line terminating at a non-finite zero, the line being at angle θ to the real axis. There will be as many asymptotes as non-finite zeros.

(*g*) The asymptotes intersect at a 'centre of gravity', each pole being considered as a 'unit positive mass' and each zero as a 'unit negative mass'. The 'centroid' will be on the real axis because the pole-zero pattern is symmetrical about the real axis. An expression for the position of the centroid is found by considering the general transfer function $K\overline{\Phi}_1$ given by equation 7.1. Equation 7.1 can be expressed as follows

$$\overline{\Phi}_a = K\overline{\Phi}_1 = \frac{K(s^n + a_{n-1}s^{n-1} + a_{n-2}s^{n-2} + \ldots a_0)}{(s^m + b_{m-1}s^{m-1} + b_{m-2}s^{m-2} + \ldots b_0)} \quad (7.17)$$

The numerator of equation 7.17 can be expressed as n roots which are the finite zeros of $\overline{\Phi}_a$. The denominator can be expressed as m roots which are the poles of $\overline{\Phi}_a$. For large values of s equation 7.17 approximates to

$$K\overline{\Phi}_1 = \frac{Ks^n}{s^m}$$

therefore

$$\overline{\Phi}_1 = \frac{1}{s^{m-n}} \quad (7.18)$$

The condition for the point s to be on a root-locus is $K\overline{\Phi}_1 + 1 = 0$ and equation 7.18 becomes

$$s^{m-n} = -K \quad (7.19)$$

We can see that equation 7.19 must, in fact, be the equation for the asymptotes of the root-loci terminating on $(m-n)$ non-finite zeros. Equation 7.19 represents $(m-n)$ roots at a distance K from a central point; the roots are symmetrical about the central point. (Students can verify this by considering a specific example, say $m-n = 3$. The three values of s which satisfy equation 7.19 are then $(\frac{1}{2}+j\sqrt{3}/2)\,K^{1/3}$, $(\frac{1}{2}-j\sqrt{3}/2)\,K^{1/3}$ and $-K^{1/3}$.) As K is varied, the roots only vary in modulus and hence represent the asymptotes. The centroid is the intersection of the asymptotes. At a large value of s the root pattern given by equation 7.19 will be a true pattern for the system and must have the same centroid as the pole-zero pattern of the system. The effect is as if all roots were located at the centroid. The centroid of the pole-zero pattern is found by taking moments of the real part about the origin (the imaginary parts of the conjugate complex roots cancel). As all roots of the equation $s^{m-n} = -K$ are considered to be located at the centroid and there are $(m-n)$ roots, the moment of the pattern is $(m-n)d$, where d is the position of the centroid along the real axis. This moment is equated to the moment of the real parts of the pole-zero pattern about the origin,

$$\underbrace{(m-n)d}_{\substack{\text{(equal roots}\\ \text{located at}\\ \text{the centroid)}}} = \underbrace{\Sigma \text{ real part of the poles}}_{\text{(pole is a 'unit positive mass')}} - \underbrace{\Sigma \text{ real part of the zeros}}_{\substack{\text{(zero is a 'unit}\\ \text{negative mass')}}}$$

therefore

$$d = \frac{\Sigma \text{ poles} - \Sigma \text{ zeros}}{m-n} \tag{7.20}$$

$(m-n)$ represents the number of non-finite zeros.

(*h*) In note (*e*) we found a root-locus on the real axis going between two poles. As we expect loci to terminate on zeros, this cannot be the complete locus. A locus of this kind breaks away from the real axis at a point. The locus will break away in opposite directions and at 90° to the real axis. If we examine a $|\overline{\Phi}|$-surface, which has this condition (figure 7.8), we see that it occurs when the $|\overline{\Phi}|$-surface is a minimum. Therefore, the break-away point(s) can be found by determining $d\overline{\Phi}_a/ds$ and equating to zero.

(*i*) It is a help to the construction of the root-locus pattern if we know whether it intersects the imaginary axis. If ω is the value of intersection, it is found by substituting $s = j\omega$ into the root-locus equation, $1+K\overline{\Phi}_1 = 0$. Equating imaginary parts of the resulting equation to zero will give the required value of ω.

The preceding notes can be expressed as rules for constructing root-locus patterns.

Rule (*i*) *Pole-zero pattern*

The n finite zeros and the m poles are plotted on the s-plane. $(m-n)$ indicates the number of non-finite zeros.

Rule (ii) Number of root-loci

The number of poles gives the number of loci.

Rule (iii) Root-loci along the real axis

A value of s on the real axis is a point on a root-locus if the total number of poles and zeros on the real axis to the right of the point is odd.

Rule (iv) Break-away from the real axis

The break-away point of a root-locus from the real axis is found by differentiating the open-loop transfer function with respect to s and equating to zero.

Rule (v) Asymptotes to root-loci at infinity

There are as many asymptotes as non-finite zeros. The angle(s) of the asymptotes are given by

$$\theta = \frac{180^\circ + N360^\circ}{n-m}$$

where $(n-m)$ is a negative number representing the number of non-finite zeros.

Rule (vi) Intersection of asymptotes

The asymptotes intersect at a centroid on the real axis. The position of the centroid is given by

$$d = \frac{\Sigma \text{ poles} - \Sigma \text{ zeros}}{m-n}$$

In summing the poles and zeros, only the real parts need be considered as the complex roots are always conjugate complex pairs.

Rule (vii) Intersection with the imaginary axis

Intersection of a root-locus with the imaginary axis is found by substituting $j\omega$ for s and equating imaginary parts to zero to find ω, the intersection.

Rule (viii) Angles of departure and arrival

The angles at which the loci depart from the poles and arrive at the finite zeros is obtained using:

angle of departure from pole q = Σ angles of the vectors from the zeros to $q - \Sigma$ angles of the vectors from the poles to $q - (180^\circ + N360^\circ)$ (7.21)

The angle of arrival at a finite zero is obtained from the same equation, except that the sign of the right-hand side must be changed.

N.B. To find points on a root-locus the following relationship must be satisfied:

$$\Sigma \angle Z - \Sigma \angle P = 180° + N360°$$

where $\Sigma \angle Z$ is the sum of the phase angles of the vectors from the zeros to the point under consideration and $\Sigma \angle P$ is the sum of the phase angles from the poles. The angles are measured in an anti-clockwise direction from the $+\alpha$-axis.

Before using the root-locus pattern in relation to control systems we will plot some root-locus patterns to illustrate the rules we have developed.

Examples

1. *Root-locus pattern of a first-order system*

We shall find the root-locus pattern of the system whose forward path transfer function is

$$\bar{\Phi}_a = \frac{20}{s0{\cdot}1+1}$$

$$K\bar{\Phi}_1 = 200\left(\frac{1}{s+10}\right)$$

$$K = 200 \quad \text{and} \quad \bar{\Phi}_1 = \frac{1}{s+10}$$

Rule (*i*) Pole-zero pattern

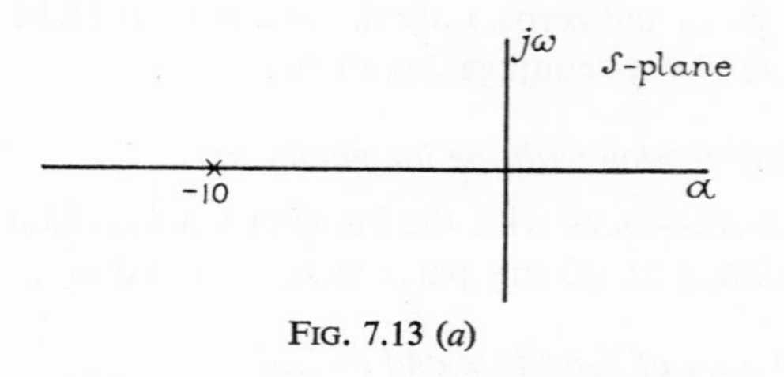

FIG. 7.13 (*a*)

Figure 7.13(*a*). There is one pole at $s = -10$ and one non-finite zero. $(n = 0,\ m = 1)$

Rule (*ii*) Number of root-loci.

There is one pole, therefore one root-locus terminating at a non-finite zero.

Rule (*iii*) Root-loci along the real axis

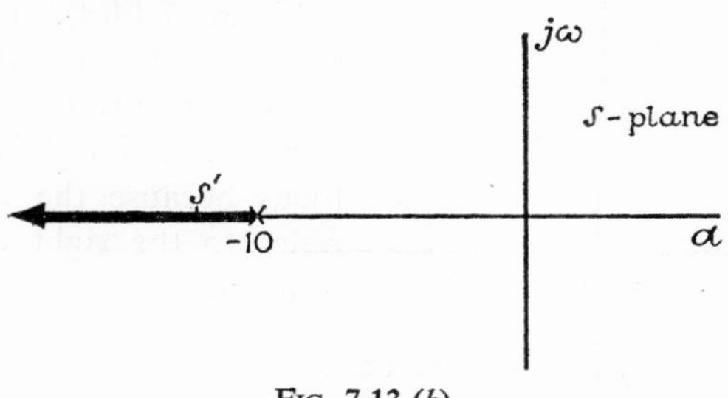

FIG. 7.13 (*b*)

Figure 7.13(*b*). Point s' is on the root-locus because the total number of poles to the right of it is an odd number.

The root-locus pattern for the first-order system is as shown in figure 7.13(*b*).

2. *Root-locus pattern of a second-order system*

We shall find the root-locus pattern of the system whose forward path transfer function is

$$\overline{\Phi}_a = \frac{K}{(s+2)(s+3)}$$

Rule (*i*) Pole-zero pattern

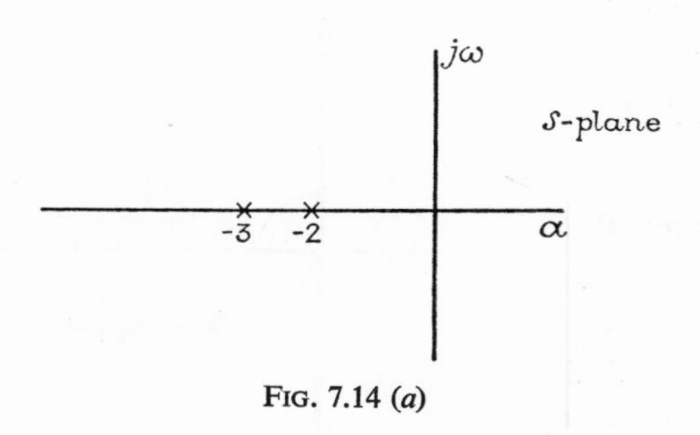

FIG. 7.14 (*a*)

Figure 7.14(*a*). There are two poles at $s = -2$ and at $s = -3$ respectively, and there are two non-finite zeros. ($n = 0$, $m = 2$)

Rule (*ii*) Number of root-loci.

There are two poles, therefore two root-loci, both terminating at non-finite zeros.

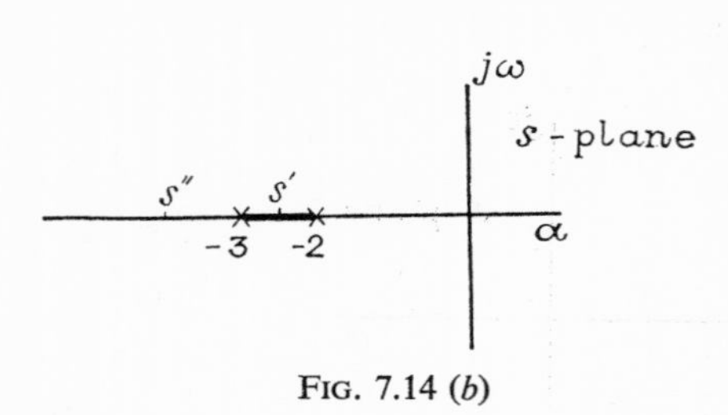

FIG. 7.14 (*b*)

Rule (*iii*) Root-loci along the real axis

Figure 7.14(*b*). Point s' is on a root-locus because the total number of poles to the right of it is an odd number. Point s'' is not on a root-locus because the total number of poles to the right of it is an even number.

Rule (*iv*) Break-away from the real axis

There are two root-loci and therefore the loci on the real axis must break away at a point. The break-away point is obtained from

$$\frac{d\Phi_a}{ds} = 0$$

therefore

$$2s+5 = 0$$

$$s = -2{\cdot}5$$

The break-away point is $-2{\cdot}5$.

Rule (*v*) Asymptotes to the root-loci at infinity.

There are two non-finite zeros and hence two asymptotes. The angle of the symptotes to the real axis is given by

$$\theta = \frac{180° + N360°}{n-m}$$

$$\theta = \frac{180° + N360°}{-2}$$

$$= 270° \text{ and } 90°$$

There are two asymptotes at 90° and 270° to the real axis.

Figure 7.14(*c*). This figure shows the complete root-locus pattern. The asymptotes in this example are part of the root-loci.

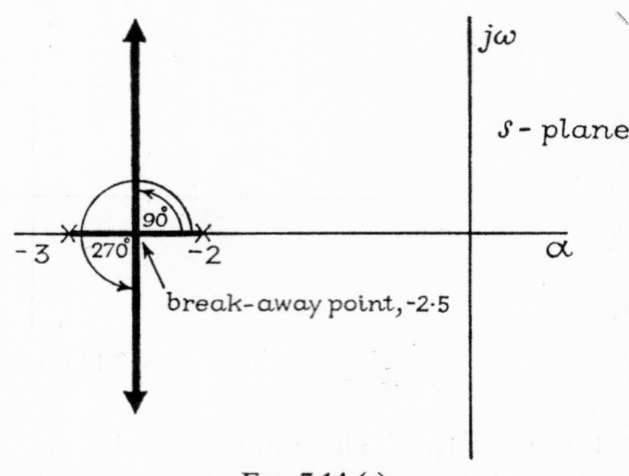

FIG. 7.14 (*c*)

3. *Root-locus pattern of a third-order system*

We shall find the root-locus pattern of the system whose forward path transfer function is

$$\overline{\Phi}_a = \frac{K}{s(s+2)(s+3)}$$

Rule (*i*) Pole-zero pattern

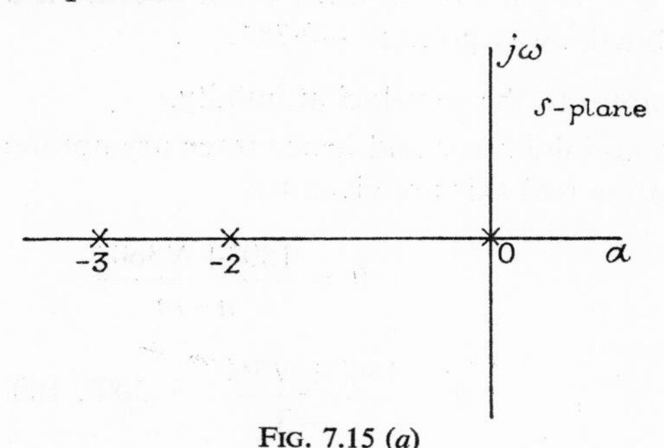

FIG. 7.15 (*a*)

Figure 7.15(*a*). There are three poles at $s = 0$, $s = -2$ and $s = -3$ respectively, and there are three non-finite zeros. ($n = 0$, $m = 3$)

Rule (*ii*) Number of root-loci.

There are three poles, therefore three root-loci, all terminating at non-finite zeros.

Rule (*iii*) Root-loci along the real axis

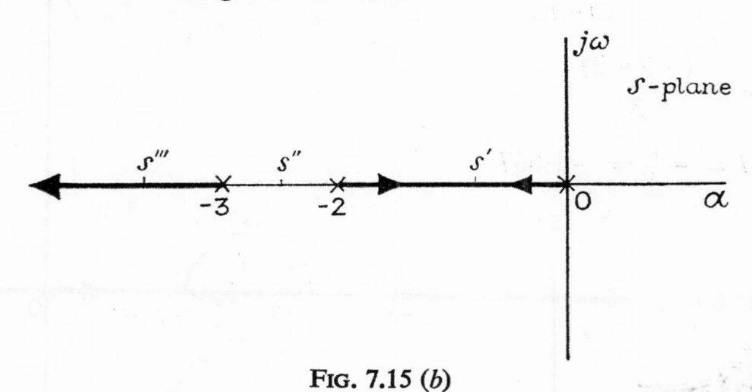

FIG. 7.15 (*b*)

Figure 7.15(*b*). Points s' and s''' are on root-loci, because the total number of poles to the right of them is an odd number. Point s'' is not on a root-locus because the total number of poles to the right of it is an even number.

Rule (*iv*) Break-away from the real axis.

There is a break-away point between the poles $s = 0$ and $s = -2$. The break-away point is obtained from

$$\frac{d\overline{\Phi}_a}{ds} = 0$$

therefore, $$3s^2+10s+6 = 0$$

therefore $$s = -2{\cdot}549 \text{ or } -0{\cdot}784$$

$s = -2{\cdot}549$ is not a possible break-away point because it does not lie on a root-locus. The break-away point is $-0{\cdot}784$.

Rule (*v*) Asymptotes to the root-loci at infinity.

There are three non-finite zeros and hence three asymptotes. The angles of the asymptotes to the real axis are given by

$$\theta = \frac{180°+N360°}{n-m}$$

$$\theta = \frac{180°+N360°}{-3} = 300°, 180° \text{ and } 60°$$

There are three asymptotes at 60°, 180° and 300° to the real axis.

Rule (*vi*) Intersection of the asymptotes.

The three asymptotes intersect on the real axis at a centroid given by

$$d = \frac{\Sigma \text{ poles} - \Sigma \text{ zeros}}{m-n}$$

$$d = \frac{(-2-3)}{3}$$

therefore $d = -\frac{5}{3} = -1{\cdot}667$

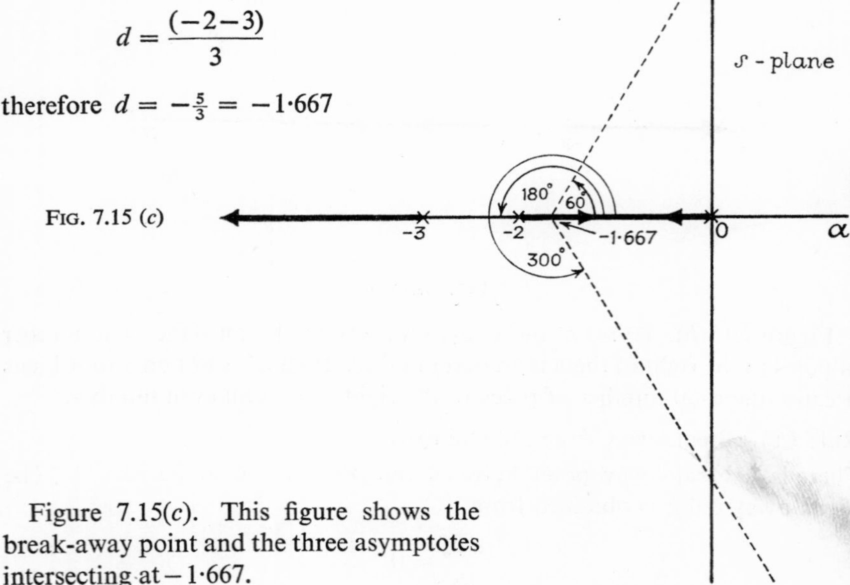

FIG. 7.15 (*c*)

Figure 7.15(*c*). This figure shows the break-away point and the three asymptotes intersecting at $-1{\cdot}667$.

Rule (*vii*) Intersection with the imaginary axis.

Intersection with the imaginary axis is found by substituting $j\omega$ for s into $\overline{\Phi}_a$ and equating the imaginary parts to zero.

$$[\overline{\Phi}_a]_{s=j\omega} = \frac{K}{j\omega(j\omega+2)(j\omega+3)}$$

$$[\overline{\Phi}_a]_{s=j\omega} = \frac{K(-5\omega^2 - j(6\omega-\omega^3))}{25\omega^4+(6\omega-\omega^3)^2}$$

Equating the imaginary parts to zero we have

$$0 = 6\omega - \omega^3$$

therefore $\omega = \pm 2{\cdot}449.$

The root-loci intersect the imaginary axis at $+2{\cdot}449$ and $-2{\cdot}449$.

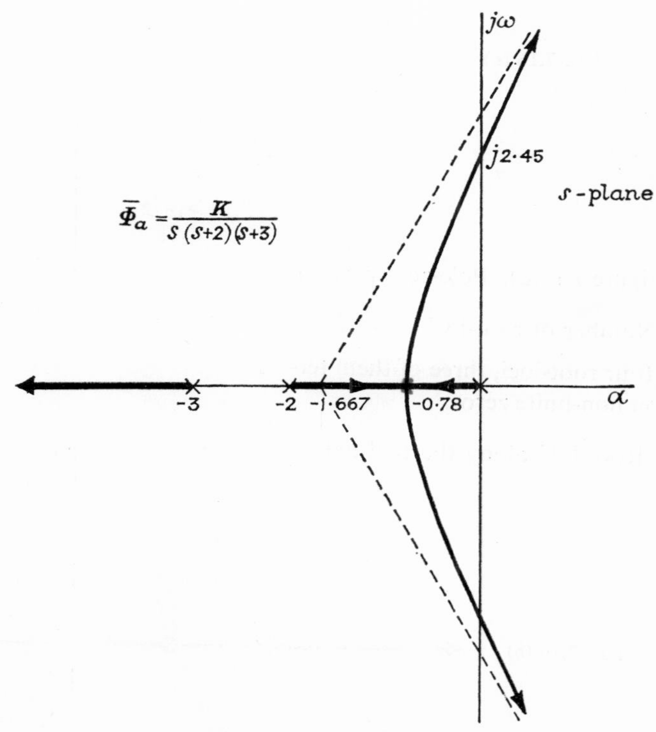

FIG. 7.15 (*d*)

Figure 7·15(*d*). The complete root-locus pattern is shown in this figure. The shape of the pattern has been sketched in with the available points.

Further points can be checked by reading them off the sketch and using equation 7.14 to verify whether they are accurately on the root-locus.

4. *Root-locus pattern of a fourth-order system*

Students should now be familiar with the construction rules and the rule number and title will just be stated in constructing this root-locus pattern.

We shall find the root-locus pattern of the system whose forward path transfer function is

$$\overline{\Phi}_a = \frac{K(s+1)}{s(s+2)(s^2+2s+5)}$$

$$\overline{\Phi}_a = \frac{K(s+1)}{s(s+2)(s+1+j2)(s+1-j2)}$$

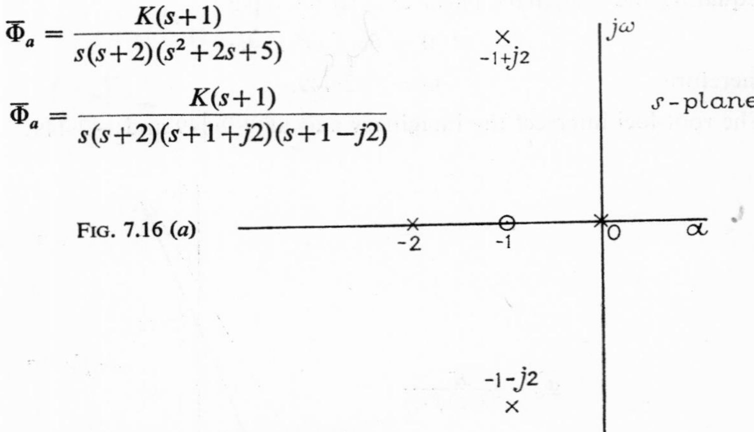

FIG. 7.16 (*a*)

Rule (*i*) Figure 7.16(*a*). Pole-zero pattern.

Rule (*ii*) Number of root-loci.

There are four root-loci, three of them terminating on non-finite zeros.

Rule (*iii*) Root-loci along the real axis.

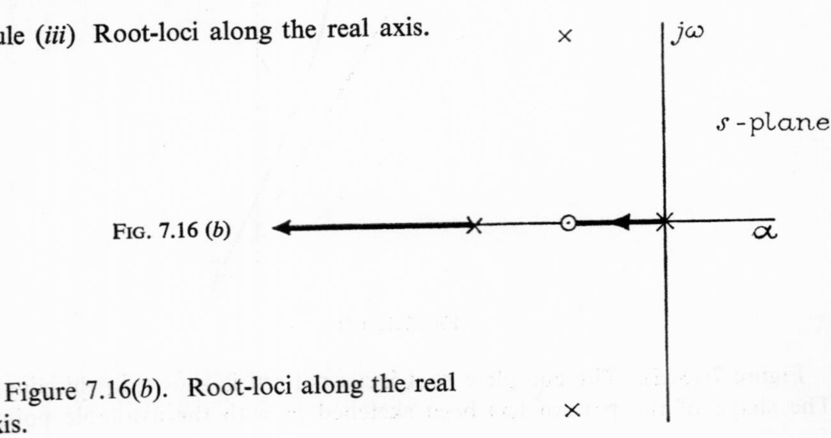

FIG. 7.16 (*b*)

Figure 7.16(*b*). Root-loci along the real axis.

Rule (*v*) Asymptotes to the root-loci at infinity

$$\theta = \frac{180° + N360°}{n-m}$$

$$\theta = \frac{180° + N360°}{-3} = 300°,\ 180° \text{ and } 60°$$

There are three asymptotes at 60°, 180° and 300° to the real axis.

Rule (*vi*) Intersection of the asymptotes

$$d = \frac{\Sigma\ \text{poles} - \Sigma\ \text{zeros}}{m-n}$$

$$d = \frac{(-2-1+j2-1-j2)}{3} \quad - (-1)$$

$$d = -1$$

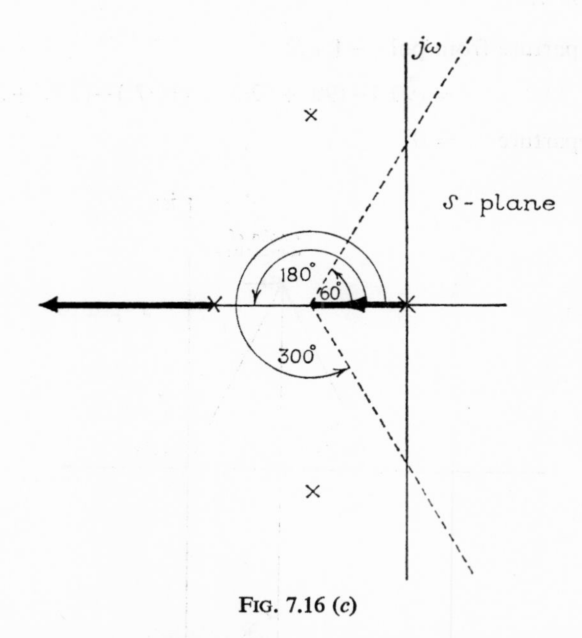

FIG. 7.16 (*c*)

Figure 7.16(*c*). This figure shows the three asymptotes intersecting at −1.

Rule (*vii*) Intersection with the imaginary axis

$$[\overline{\Phi}_a]_{s\,=\,j\omega} = \frac{K(1+j\omega)}{\omega^4 - 9\omega^2 + j(10\omega - 4\omega^3)}$$

Equating imaginary parts to zero, we obtain

$$0 = j\omega^5 - 9j\omega^3 - 10j\omega + 4j\omega^3$$

therefore, $$\omega^2 = 6{\cdot}53 \text{ or } -1{\cdot}53$$

As the 'number' ω must be real $-1{\cdot}53$ is not a result, therefore

$$\omega^2 = 6{\cdot}53$$

$$\omega = \pm 2{\cdot}56$$

The root-loci intersect the imaginary axis at $+2{\cdot}56$ and $-2{\cdot}56$.

Rule (*viii*) Angles of departure and arrival.

We must know the angle of departure of the root-loci from the conjugate complex poles to obtain the general shape of the root-locus.
Using equation 7.21,

angle of departure from pole $-1+j2$

$$= (90°) - (90° + 63{\cdot}3° + 116{\cdot}7°) - (180° + N360°)$$

angle of departure $= 0°$

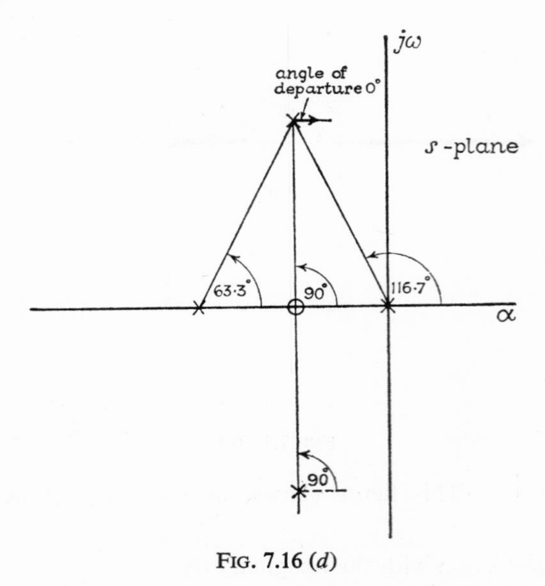

FIG. 7.16 (*d*)

Figure 7.16(*d*). The angle of departure from the pole at $-1+j2$ is 0°. The angles used in the calculations can be measured from the *s*-plane sketch with a protractor. As the pole-zero pattern is symmetrical the angle of departure from the pole $-1-j2$ is the 'mirror image' of the angle of departure from the pole $-1+j2$.

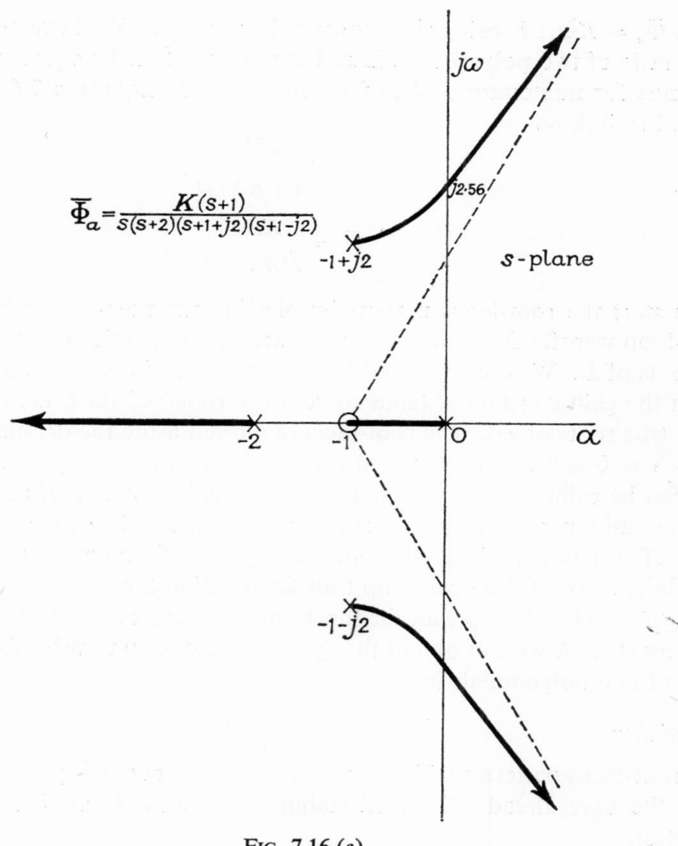

FIG. 7.16 (*e*)

Figure 7.16(*e*). This figure shows a sketch of the root-locus pattern of the system using the information we have obtained by applying the construction rules.

7.3 Root-locus methods applied to control systems

We are now in a position to obtain the root-locus pattern of a control system. In this section we shall examine the root-locus patterns of systems and develop various ideas on the interpretation of the patterns. Let us reconsider the equation of the closed-loop transfer function $\overline{\Phi}$ of a single-loop control system (p. 13).

$$\overline{\Phi} = \frac{\overline{\Phi}_a}{1+\overline{\Phi}_a} \tag{7.22}$$

where $\overline{\Phi}_a = K\overline{\Phi}_1$, K being the system gain constant. We have seen that $\overline{\Phi}_1$ is the ratio of two polynomials in s. Let the ratio $\overline{N}/\overline{D}$ be equal to $\overline{\Phi}_1$ where N stands for numerator and D for denominator. Equation 7.6 can be expressed as follows

$$\overline{\Phi} = \frac{K\overline{N}/\overline{D}}{1+K\overline{N}/\overline{D}}$$

$$\overline{\Phi} = \frac{K\overline{N}}{\overline{D}+K\overline{N}} \tag{7.23}$$

We start the root-locus pattern by plotting the pole-zero pattern of the open-loop transfer function. The zeros are the roots of N and the poles are the roots of D. We are interested in the behaviour of the closed-loop system. When the gain constant K tends to zero the root-loci start, i.e. at the poles of $\overline{\Phi}_1$ (the roots of D). The root-loci are plotted using the defining equation $\overline{D}+K\overline{N} = 0$ and hence, represent a plot of the *closed-loop poles*. The root-loci can be calibrated for different values of gain constant K and represent the closed-loop poles at any particular value of K. Thus, we can read the roots of equation 7.23 off the root-locus pattern for any given value of K. The finite zeros of the open-loop transfer function are the same as the finite zeros of the closed-loop transfer function. We have established that for any gain constant K we can obtain the system closed-loop transfer function as a ratio of two polynomials in s.

Stability

The root-locus pattern enables us to see whether a control system is stable or not. We have already discussed stability in general terms (chapter 4, section 4.2).

To see how the root-locus pattern indicates whether a system is unstable consider the closed-loop transfer function in general partial fraction form,

$$\overline{\Phi} = K\overline{N}\left(\frac{A_1}{s-q_1}+\frac{A_2}{s-q_2}+\ldots+\frac{A_n}{s-q_n}\right) \tag{7.24}$$

In equation 7.24, $q_1, q_2, q_3, \ldots q_n$ are the closed-loop poles and $A_1/(s-q_1)$, $A_2/(s-q_2), \ldots A_n/(s-q_n)$ are the partial fractions. The time solution can be obtained by inverse transformation. If conjugate complex roots lie on the right-hand side of the s-plane they indicate instability. This may be verified as follows:

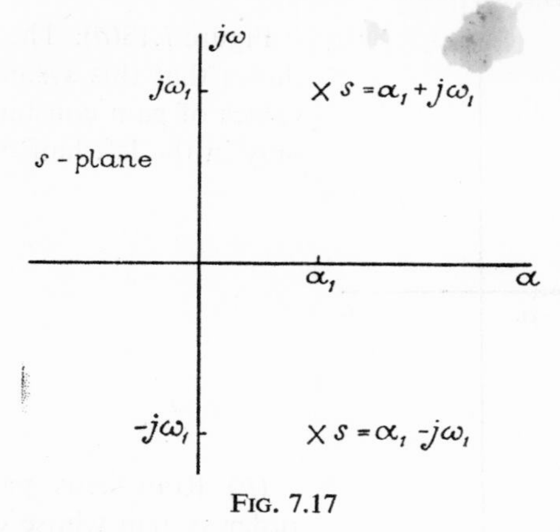

FIG. 7.17

Figure 7.17. Conjugate complex roots $\alpha_1 + j\omega_1$ and $\alpha_1 - j\omega_1$ are shown on the right half of the s-plane. If these roots are q_1 and q_2 in the general equation 7.24, the first two terms of the equation are

$$\frac{A_1}{s-(\alpha_1+j\omega_1)}+\frac{A_2}{s-(\alpha_1-j\omega_1)}$$

The inverse transform of the above partial fractions is

$$A_1\,\mathrm{e}^{\alpha_1 t}\mathrm{e}^{j\omega_1 t}+A_2\,\mathrm{e}^{\alpha_1 t}\mathrm{e}^{-j\omega_1 t} = \mathrm{e}^{\alpha_1 t}\,(A\sin\omega_1 t+B\cos\omega_1 t) \tag{7.25}$$

where A and B are constants. Expression 7.25 indicates an increasing oscillatory effect with time t, which is obviously undesirable at the output of a system. Any increasing exponential is a source of instability and hence, for a stable system we can have no poles appearing in the right half of the s-plane. A value of gain constant K that 'takes' the root-locus pattern into the right half of the s-plane indicates an unstable system. If conjugate complex roots lie on the imaginary axis, the time solution for them is of the form $A\sin\omega t+B\cos\omega t$. This will give a constant amplitude oscillation at the output, which is undesirable. We can conclude that for a system to be stable the roots must lie in the left half of the s-plane. We shall now look, with reference to the system gain constant, at three root-locus patterns from the point of view of stability.

(*i*) Root-locus pattern of a second-order system whose open-loop transfer function is

$$\Phi_a = \frac{K}{(s+a)(s+b)}$$

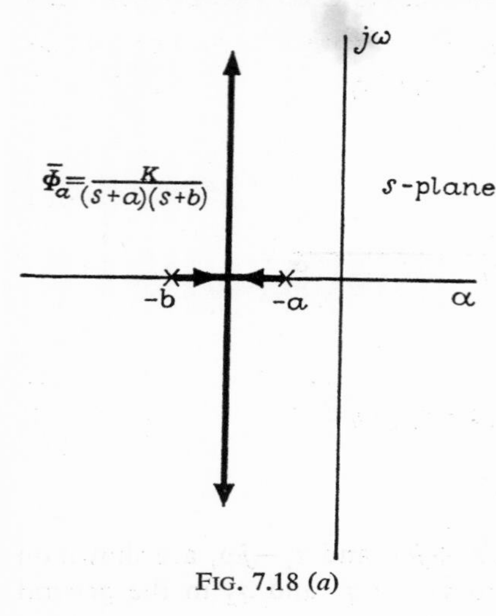

FIG. 7.18 (*a*)

Figure 7.18(*a*). The root-locus pattern shows that this system is stable for all values of gain constant *K*, because it is only in the left half of the *s*-plane.

(*ii*) Root-locus pattern of a third-order system whose open-loop transfer function is

$$\bar{\Phi}_a = \frac{K}{s(s+a)(s+b)}$$

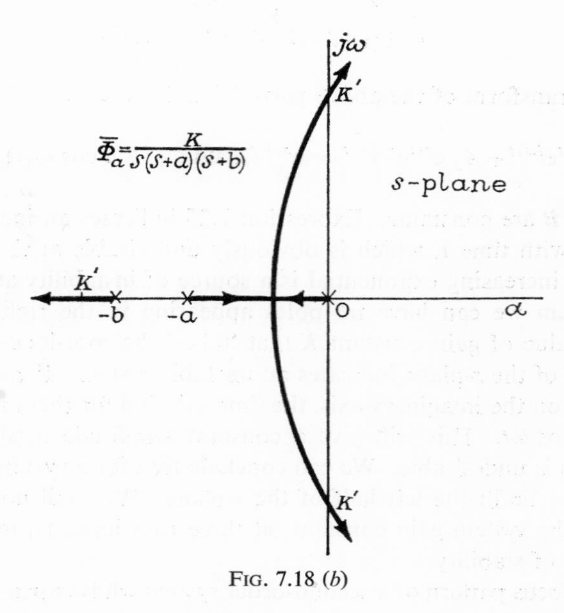

FIG. 7.18 (*b*)

Figure 7.18(*b*). The root-locus pattern is bent into the right half of the *s*-plane and for values of the gain constant greater than K' the system is unstable.

(*iii*) Root-locus pattern of a third-order system with a finite zero, whose open-loop transfer function is

$$\overline{\Phi}_a = \frac{K(s+c)}{s(s+a)(s+b)}$$

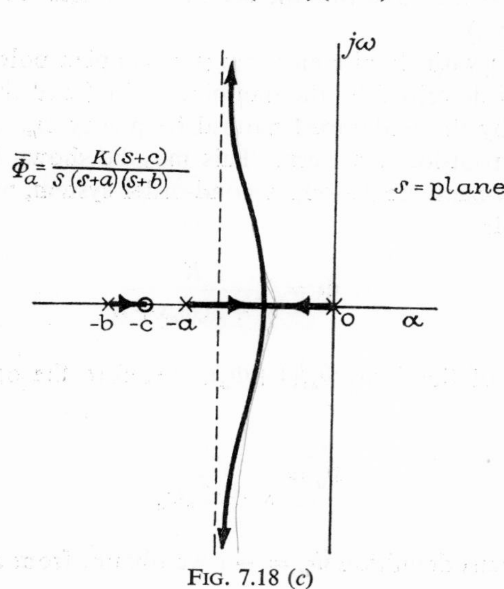

FIG. 7.18 (*c*)

Figure 7.18(*c*). The root-locus pattern is bent back into the left half of the *s*-plane and is stable for all values of gain constant.

Figures 7.18(*a*) to 7.18(*c*) show that if a pole relatively near the *jω* axis is added to the system it becomes dominant and the system becomes less stable. The addition of a zero, however, can make a system more stable.

The speed of response and the possible overshoot of a control system's output is indicated by the nearness of the roots to the imaginary axis. The poles near the imaginary axis are called the *dominant poles.* To obtain a damped system with a good response the system is usually designed so that it is underdamped (see chapter 4). This means that there are usually dominant conjugate complex poles.

7.4 Calibration of the root-locus pattern

The root-locus pattern is calibrated for the gain constant *K* by using the modulus condition, equation 7.10, i.e.

$$|\overline{\Phi}_1| = \frac{1}{K}$$

therefore,
$$K = \frac{1}{|\overline{\Phi}_1|} = \frac{|\overline{D}|}{|\overline{N}|} \tag{7.26}$$

The value of K at a point on a root-locus can be found graphically by using equation 7.26. (To facilitate this the scales of the real and imaginary axes should be the same.)

Control systems with dominant conjugate complex poles can have their degree of stability described by the damping ratio ζ and their transient response described by the undamped natural frequency ω_n. ζ and ω_n can be indicated on the root-locus pattern. This may be shown by considering a general unity feed-back single-loop second-order system, whose closed-loop transfer function is

$$\overline{\Phi} = \frac{K}{s^2+2\zeta\omega_n s+K} \tag{7.27}$$

Equation 7.27 is of the form $\overline{\Phi}_a/(1+\overline{\Phi}_a)$, therefore the open-loop transfer function is

$$\overline{\Phi}_a = \frac{K}{s^2+s2\zeta\omega_n} \tag{7.28}$$

Using the root-locus definition $\overline{\Phi}_a = -1$ we obtain, from equation 7.28,

$$\frac{K}{s^2+2\zeta\omega_n s} = -1 \tag{7.29}$$

In chapter 4 we found K to be ω_n^2; this makes equation 7.29 a quadratic in s with the roots

$$s = -\zeta\omega_n \pm j\omega_n\sqrt{(1-\zeta^2)} \tag{7.30}$$

For a particular value of ζ equation 7.30 is the equation of a straight line passing through the origin on the s-plane. The angle the straight line makes with the negative real axis is θ, where

$$\tan\theta = \frac{\omega_n\sqrt{(1-\zeta^2)}}{\zeta\omega_n}$$

therefore
$$\cos\theta = \frac{\zeta\omega_n}{\omega_n}$$

$$\cos\theta = \zeta \tag{7.31}$$

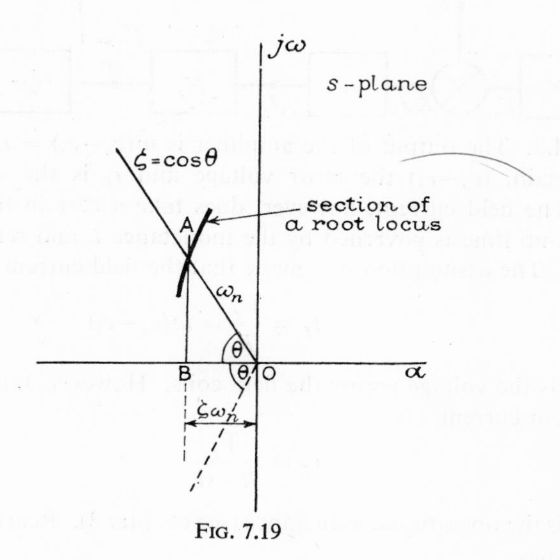

FIG. 7.19

Figure 7.19. The figure shows equation 7.30 plotted on the *s*-plane for a particular value of damping ratio ζ. Equation 7.30 really represents two lines, but as the root-locus pattern is symmetrical about the real axis we need not consider both lines. The line OA represents a damping ratio of $\zeta = \cos\theta$. The gain constant K for this damping ratio is where the line cuts the root-locus; the undamped natural frequency ω_n is the distance OA (from equation 7.31). The damping ratio line can be applied to any system with dominant conjugate complex poles.

7.5 Root-locus pattern adjustments

The shape of root-locus pattern indicates a system's performance, and the performance may be improved by changing the root-locus pattern. The shape can be changed by 'adding' poles and zeros to the open-loop transfer function. This is illustrated by the following example.

Example

The simple remote position control system discussed in chapter 4 will be used as an example of the root-locus pattern. Before we start using the root-locus pattern we shall examine the system in a little more detail. Consider the block diagram shown at the top of page 128.

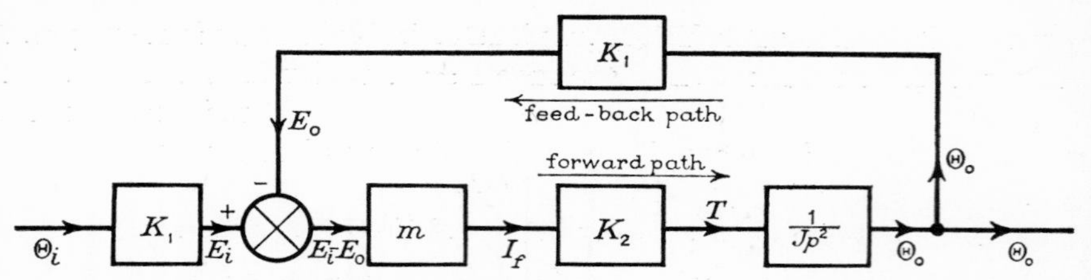

Figure 4.2. The output of the amplifier is $m(e_o - e_i) = i_f$ where m is the amplifier gain, $(e_o - e_i)$ the error voltage and i_f is the steady-state field current. The field current, however, does take a certain time to build up. This build-up time is governed by the inductance L and resistance R of the field coils. The assumption was made that the field current

$$i_f = \frac{V_f}{R} = m(e_o - e_i) \tag{7.32}$$

where V_f is the voltage across the field coils. However, taking into account the transient current

$$\bar{i}_f = \frac{\bar{V}_f}{R + sL} \tag{7.33}$$

where sL is the operational inductance (see chapter 3). Rearranging, equation 7.33 becomes

$$\bar{i}_f = \frac{\bar{V}_f}{R(1 + sL/R)} \tag{7.34}$$

where L/R is known as the field time constant T_f. Equation 7.34 indicates that, in order to obtain an accurate transfer function, equation 7.32 should be modified by a factor $1/(1 + sT_f)$ as follows

$$\frac{\bar{i}_f}{(\bar{e}_o - \bar{e}_i)} = \frac{m}{1 + sT_f} \tag{7.35}$$

Equation 7.35 suggests that the amplifier can be 'replaced' by a block of transfer function $m/(1 + sT_f)$ which takes into account the build-up time of the field current. Let us see how this alters the closed- and open-loop transfer functions of the complete system. The open-loop transfer function (equation 7.28), which is the forward path transfer function, was

$$\frac{\omega_n^2}{s^2 + 2\zeta\omega_n s}$$

It is now modified to

$$\frac{1}{T_f(s + 1/T_f)} \cdot \frac{\omega_n^2}{(s^2 + 2\zeta\omega_n s)}$$

The block diagram can be simplified as shown in figure 7.20.

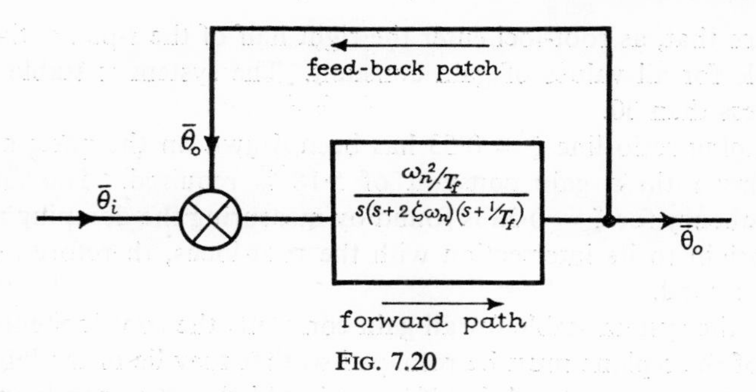

FIG. 7.20

Figure 7.20. This figure shows the block diagram for the simple remote position control system. The system is now a third-order system and its general open-loop transfer function is

$$\bar{\Phi}_a = \frac{K}{s(s+2\zeta\omega_n)(s+1/T_f)} \tag{7.36}$$

We shall now consider an example of the simple use of the root-locus pattern for a system whose open-loop transfer function is of the form of equation 7.36. Let the open-loop transfer function of a system be

$$\bar{\Phi}_a = \frac{K}{s(s+2)(s+3)}$$

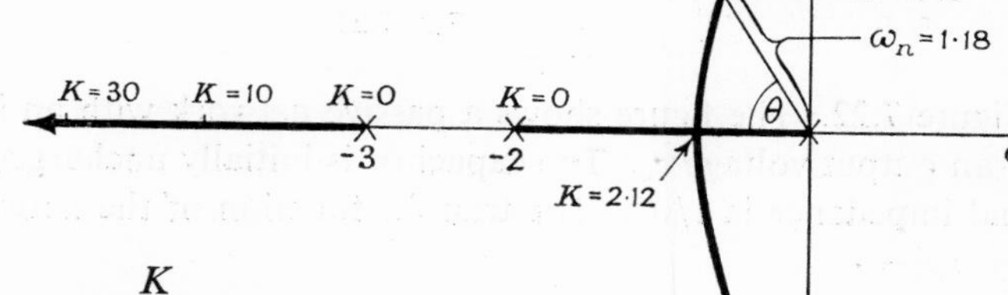

FIG. 7.21

$$\bar{\Phi}_a = \frac{K}{s(s+2)(s+3)}$$

(This root-locus pattern was plotted as the third example when we applied the root-locus pattern rules earlier in this chapter.) The root-locus pattern is shown in figure 7.21 and using equation 7.26 it has been calibrated for gain.

Figure 7.21. We can see from the root-locus pattern that, as root-loci enter the right half of the s-plane, the system is not stable for all values of gain constant. The system is stable for gain constants less than 30.

The damping ratio line $\zeta = 0{\cdot}53$ has been drawn on the pattern and for this damping ratio a gain constant of 5·18 is required. The undamped natural frequency for $\zeta = 0{\cdot}53$ is found by measuring the damping ratio line from the origin to its intersection with the root-locus, therefore $\omega_n = 1{\cdot}18$ radian per second.

To make the system stable for all gain constants the root-loci entering the right half of the s-plane must be reshaped so that they lie in the left half. It has previously been suggested that if a zero is added to the open-loop transfer function the system can be made stable for all gain constants. However, since the addition of a zero is not practical† a passive network that approximates to the addition of a zero can improve the system's performance. The network adds a pole and a zero to the transfer function, but it is arranged that the pole is in a position on the s-plane such that it has little effect on the system compared with the zero. The passive network consists of two resistors and a capacitor.

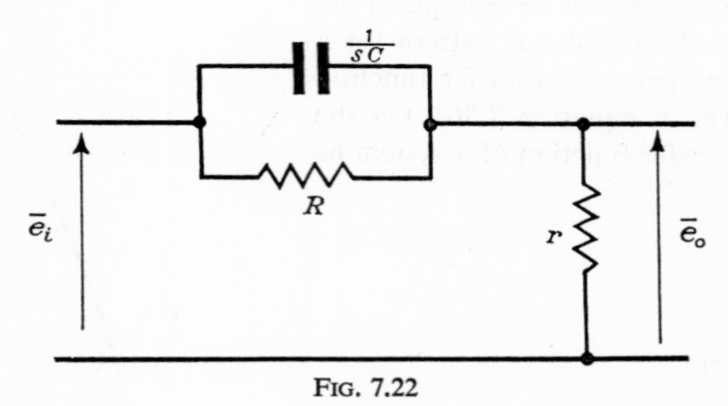

FIG. 7.22

Figure 7.22. The figure shows a passive network with an input voltage e_i and an output voltage e_o. The capacitor is initially uncharged and its operational impedance is $1/sC$. The transfer function of the network is given by

$$\frac{\bar{e}_o}{\bar{e}_i} = \frac{r}{r+R/(1+sCR)}$$

$$\frac{\bar{e}_o}{\bar{e}_i} = \frac{r(1+sCR)}{(R+r+sCRr)}$$

† Students who have read chapter 5 will realize that the addition of a zero is not easily achieved. It means the addition of a device whose transfer function is of the form $(s+a)$. If s is thought of as the operation of differentiation, the device will be some kind of differentiator which is not practical.

$$\frac{\bar{e}_o}{\bar{e}_i} = \frac{s+1/CR}{s+\dfrac{1}{CR}\cdot\dfrac{R+r}{r}} \tag{7.37}$$

Let $r/(R+r) = A$ and $CR = T$. Therefore equation 7.37 becomes

$$\frac{\bar{e}_o}{\bar{e}_i} = \frac{s+1/T}{s+1/TA} \tag{7.38}$$

Equation 7.38 is the transfer function of the passive network. If A is made small the pole $s = -1/TA$ is not very significant compared with the zero $s = -1/T$. If the network is introduced into the remote position control system before the amplifier, the root-locus pattern will be altered. The zero reshapes the important part of the root-locus, bringing it further into the left half of the s-plane, thus making the system stable for a larger range of gain constant. The network attenuates the forward path of the system and the system amplification must be increased to overcome this.

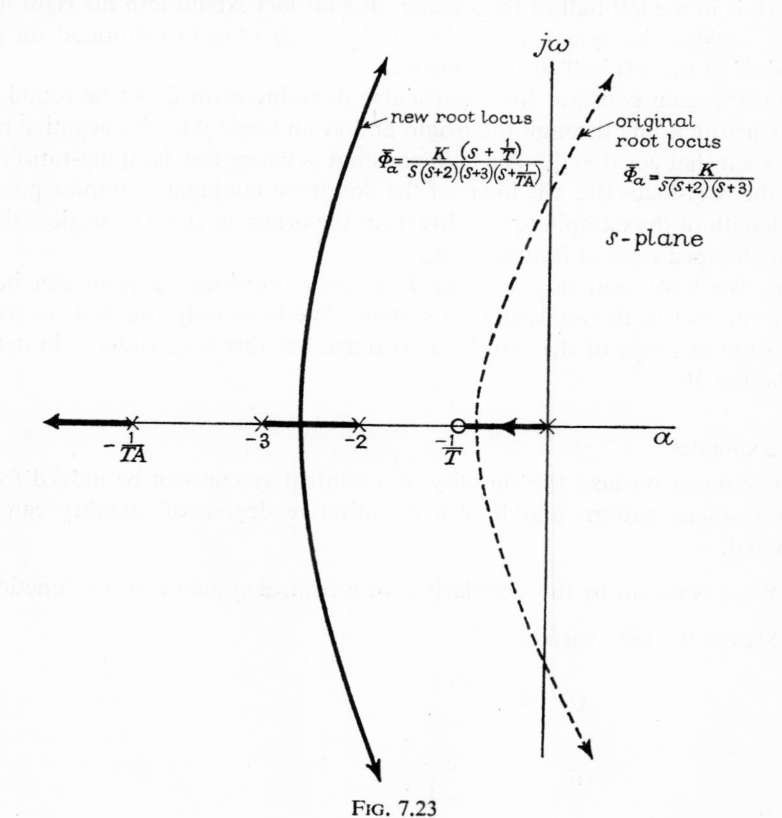

FIG. 7.23

Figure 7.23. This figure shows the root-locus pattern modified by a passive network of transfer function $(s+1/T)/(s+1/TA)$. The broken line indicates the original root-loci of the dominant conjugate complex poles. The passive network is said to provide *passive lead compensation.* It makes the system more stable and speeds up its response (see chapter 10).

7.6 Comments

(*i*) The closed-loop transfer function of a system, for a particular gain constant K, can be obtained from a root-locus pattern in the form

$$\frac{K\,[\text{product of the finite open-loop zeros (if any)}]}{[\text{product of the poles at the } K \text{ calibration on the root-loci}]}$$

(*ii*) The root-locus pattern shows the closed-loop poles of a system for any gain constant K.

(*iii*) A system is stable for all values of gain constant if its root-locus pattern is in the left half of the s-plane. If root-loci extend into the right half of the s-plane the system is stable for the range of gain calibrated on the root-loci in the left half of the s-plane.

(*iv*) The gain constant for a particular damping ratio ζ can be found by constructing a line through the origin and at an angle θ to the negative real axis, such that $\cos\theta = \zeta$. The gain constant is where the damping-ratio line cuts the root-locus (i.e. the locus of the dominant conjugate complex poles). The length of the damping-ratio line from the origin to the intersection gives the undamped natural frequency ω_n.

(*v*) We have seen how a control system's root-locus pattern can be a powerful tool in the analysis of a system. We have only touched on compensation in terms of the root-locus pattern, but this is considered in detail in chapter 10.

7.7 Examples

1. Comment on how the stability of a control system can be judged from its root-locus pattern and how a quantitative degree of stability can be obtained.

2. What is meant by the singularities of a control system transfer function?

3. Sketch the $|\overline{\Phi}|$-surfaces of

$$(i)\quad \overline{\Phi} = \frac{10}{s}$$

$$(ii)\quad \overline{\Phi} = \frac{5}{s+4}$$

4. The open-loop transfer function of a system is

$$\bar{\Phi}_a = \frac{K}{s(s+0{\cdot}1)(s+0{\cdot}2)}$$

Are the following points on the root-locus?

(*i*) $s = -j0{\cdot}2$
(*ii*) $s = +j0{\cdot}1414$
(*iii*) $s = -2+j0{\cdot}1$
(*iv*) $s = +0{\cdot}018+j0{\cdot}1$
(*v*) $s = -0{\cdot}018+j0{\cdot}1$

(No; Yes; No; No; Yes)

5. Plot the root-locus patterns of the systems whose open-loop transfer functions are

(*i*) $\bar{\Phi}_a = \frac{K}{s+4}$

(*ii*) $\bar{\Phi}_a = \frac{K}{(s+1)(s+4)}$

(*iii*) $\bar{\Phi}_a = \frac{K(s+1)}{(s+2)(s+4)}$

(*iv*) $\bar{\Phi}_a = \frac{K}{s(s+1)(s+2)}$

(*v*) $\bar{\Phi}_a = \frac{K}{(s+1)^2}$

(*vi*) $\bar{\Phi}_a = \frac{K}{s^2+2s+2}$

(*vii*) $\bar{\Phi}_a = \frac{K}{s^2+4}$

(*viii*) $\bar{\Phi}_a = \frac{K}{s^2(s+0{\cdot}1)(s+0{\cdot}2)}$

6. Give the range of gain constant K for which the systems in question 5 are stable.

(all K; all K; $0 < K < 6$; $0 < K < 6$;
all K; all K; no K; no K)

7. Sketch the root-locus patterns of

(*i*) a system stable for all gain constants,
(*ii*) a system unstable for all gain constants,
(*iii*) a system stable for a range of gain constant.

8. The system in question 4 has the point $s = 0{\cdot}035+j0{\cdot}05$ on its root-locus. If this point is a closed-loop pole, find the system damping ratio.

(0·573)

9. A remote position control system has the closed-loop transfer function

$$\bar{\Phi} = \frac{9}{s^2+6s+9}$$

find the following,

(*i*) the open-loop transfer function,
(*ii*) the system gain constant,
(*iii*) the undamped natural frequency,
(*iv*) the damping ratio.

Plot the root-locus pattern for any gain constant and comment on the system stability. Find the value of the gain constant for the system to have damping ratios of 0·4, 0·5 and 0·6.

(56·25; 36; 25)

10. The block diagram shows a single-loop unity feed-back system, where K_1 and K_2 are constants.

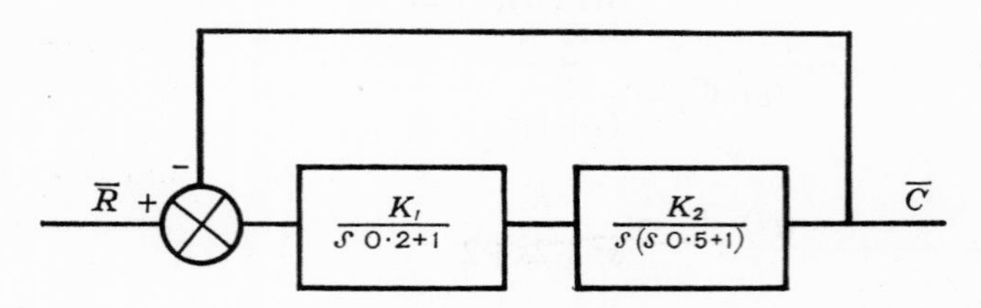

What are the open-loop and closed-loop transfer functions of the system? Sketch the root-locus pattern of the system.

11. Discuss the significance of the root-locus pattern as applied to control systems. Comment on system compensation in your discussion.

CHAPTER 8

Frequency Response

8.1 Introduction

A linear control system has a transfer function which describes the system completely. We have seen in the previous chapter how root-locus patterns can be interpreted. The root-locus method stresses the transient response of a system (i.e. it is based on the system characteristic equation, $1+\overline{\Phi}_a = 0$). Root-locus methods are sometimes called *transient response methods*.

We have already discussed the impulse response of a system which gives an easily recognizable 'physical' system response. We can readily appreciate that a time solution, i.e. a response in a time 'domain', gives a straight-forward picture of a system's behaviour. However, it is not always easy to obtain the time solution of a system.

The alternative to a time solution approach is the *frequency response*. If any time solution is periodic, it is possible to analyse it into sinusoidal components by means of Fourier analysis. It can also be shown that a non-periodic time solution can be analysed into sinusoidal components. A unit impulse contains all frequency components from 0 to ∞ hertz. If we apply sinusoidal components of frequencies from 0 to ∞ hertz to the input of a system, the sum of the steady-state outputs will be the impulse response of the system. It is important to realize that we can define the time response of a system by consideration of its *steady-state frequency response*.

We shall now consider how the frequency response can be developed from a system's differential equation (transfer function). Let us take the example of a series *RLC* circuit. The input to the circuit is the voltage e and the output is the current i. The input voltage is applied at time $t = 0$, all the initial conditions being zero.

Figure 8.1. The circuit can be described by three equations:

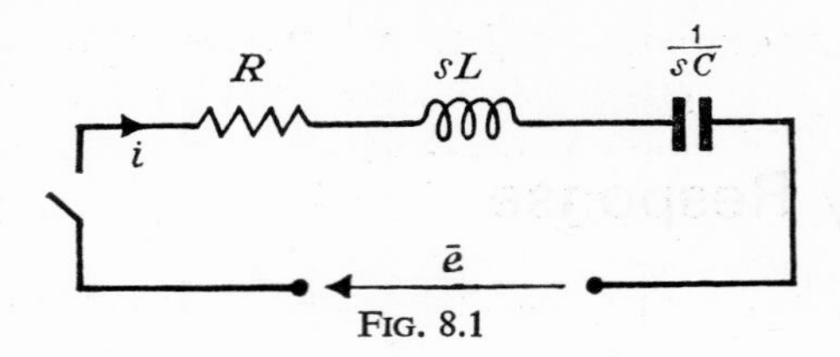

FIG. 8.1

$$e = L\frac{di}{dt}+iR+\frac{1}{C}\int_0^t i\,dt \tag{8.1}$$

$$\bar{e} = sL\bar{i}+\bar{i}R+\bar{i}/sC \tag{8.2}$$

$$e = j\omega Li+iR+i/j\omega C \tag{8.3}$$

Equations 8.1 and 8.2 are the differential equations of the circuit giving the transient and steady-state solutions. Equation 8.3 is the steady-state solution for the frequency $\omega/2\pi$ hertz. This equation gives complete information for a range of $\omega/2\pi = 0$ to $\omega/2\pi = \infty$ hertz. Then for frequency response, comparing equations 8.2 and 8.3, we have effectively replaced s by $j\omega$.

A general transfer function $\bar{\Phi}$ can be expressed as a function of frequency $\Phi(j\omega)$ by replacing s by $j\omega$. Equation 7.1 becomes

$$\Phi(j\omega) = \frac{K(j\omega-r_1)(j\omega-r_2)(j\omega-r_3)\ldots(j\omega-r_n)}{(j\omega-q_1)(j\omega-q_2)(j\omega-q_3)\ldots(j\omega-q_m)} \tag{8.4}$$

Equation 8.4 is the general transfer function of a system when considering frequency response.

The single-loop unity feed-back control system can now have an open-loop transfer function of $\Phi_a(j\omega)$, and the closed-loop transfer function becomes

$$\Phi(j\omega) = \frac{\Phi_a(j\omega)}{1+\Phi_a(j\omega)} \tag{8.5}$$

Equation 8.5 is of the same form as equation 7.6. The closed-loop transfer function $\Phi(j\omega)$ has a modulus and argument for every value of ω.

8.2 Nyquist diagrams and the Nyquist stability criterion

The frequency response from the Nyquist diagram is best approached from the point of view of stability, with reference to the s-plane.

Let us suppose that we have a contour drawn on the s-plane. This contour

or locus specifies a set of values for s. Now if we were to examine the characteristic equation, $1+\overline{\Phi}_a = 0$, for all of these values of s we would obtain a series of values with modulus and argument. If we plot the values of $1+\overline{\Phi}_a$, for each value of s on the locus, onto a plane with real and imaginary axes, we obtain another locus. This is a locus of $1+\overline{\Phi}_a$ and the plane can be called the $(1+\overline{\Phi}_a)$-plane. The transfer of points from one plane to another is called *conformal transformation.*

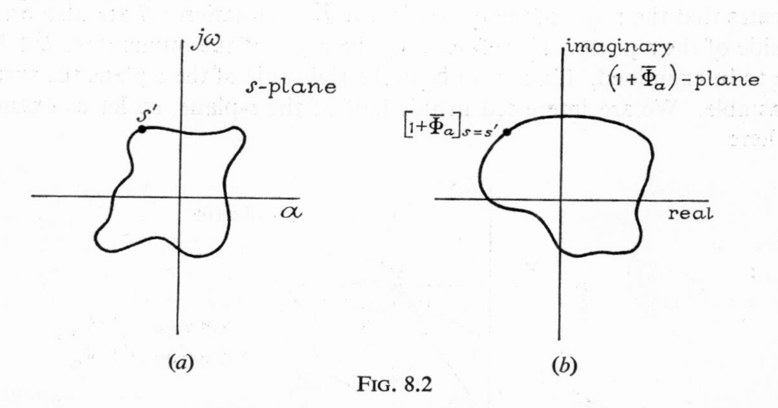

FIG. 8.2

Figures 8.2(a) and (b) show a locus transformed from the s-plane to the $(1+\overline{\Phi}_a)$-plane. The transfer for point s' on the s-plane is achieved by substituting s' into $1+\overline{\Phi}_a$ and plotting the resulting modulus and argument on the $(1+\overline{\Phi}_a)$-plane. In general a closed curve transfers into another closed curve.

Before we proceed further let us reconsider the criterion for stability with the s-plane. In plotting the root-locus pattern, i.e. the path of the roots of the characteristic equation $1+\overline{\Phi}_a = 0$, it was found that if the locus entered the right half of the s-plane the system could be unstable. Thus we concluded that if a system has roots of the characteristic equation in the right half of the s-plane the system is unstable. We are in a position to place two conditions on the systems under consideration. The first one is that the open-loop transfer function is of the form of equation 7.1, the order of the numerator being less than that of the denominator ($n < m$). The second one is that the feed-forward elements are stable in themselves (the transfer function $\overline{\Phi}_a$ represents a stable system). These conditions are true for systems within our scope.

We can rewrite the characteristic equation as follows,

$$1+K\frac{\overline{N}}{\overline{D}} = 0 \tag{8.6}$$

where $\overline{\Phi}_a = K\overline{N}/\overline{D}$, K being the gain constant and N and D standing for numerator and denominator respectively. Equation 8.6 can be rearranged to give

$$\frac{\bar{D}+K\bar{N}}{\bar{D}} = 0 \tag{8.7}$$

The second condition imposed on the systems is that $\bar{\Phi}_a = K\bar{N}/\bar{D}$ is stable, which means that all its roots are on the left-hand side of the s-plane. This indicates that the roots of the denominator $\bar{D}$ of equation 8.7 are also on the left side of the s-plane. Therefore, only the roots of the numerator, $\bar{D}+K\bar{N}$, have to be examined. If any root is on the right side of the s-plane the system is unstable. We are interested in this half of the s-plane, so let us examine loci here.

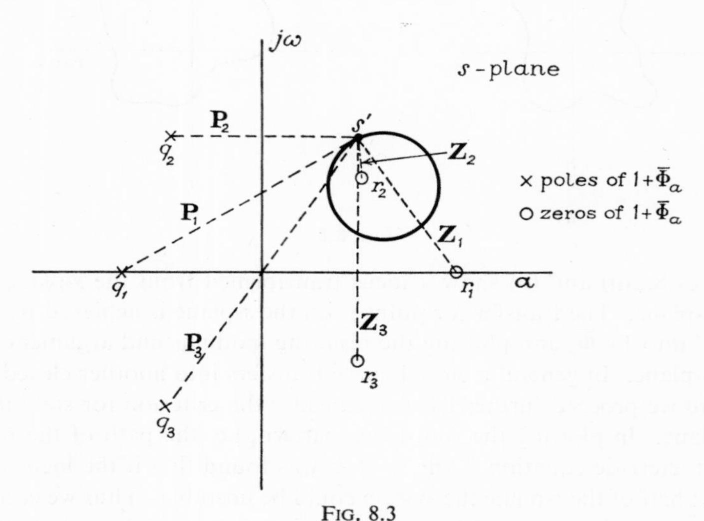

FIG. 8.3

Figure 8.3. This figure shows a circular locus on the right side of the s-plane. q_1, q_2 and q_3 are the poles of a particular $1+\bar{\Phi}_a$ (as $\bar{\Phi}_a$ is stable they are in the left half of the s-plane). r_1, r_2 and r_3 are the zeros of $1+\bar{\Phi}_a$. $\mathbf{P}_1$, $\mathbf{P}_2$, $\mathbf{P}_3$ and $\mathbf{Z}_1$, $\mathbf{Z}_2$, $\mathbf{Z}_3$ are the vectors from the poles and zeros to a point s' on the circular locus. We can transform the circular locus onto the $(1+\bar{\Phi}_a)$-plane by plotting the modulus and argument of $1+\bar{\Phi}_a$. When $s = s'$, $1+\bar{\Phi}_a = (\mathbf{Z}_1\mathbf{Z}_2\mathbf{Z}_3)/(\mathbf{P}_1\mathbf{P}_2\mathbf{P}_3)$. However, we only need to consider the *phase change* of $1+\bar{\Phi}_a$ in moving point s' in a clockwise direction around the circular locus. The phase change of $1+\bar{\Phi}_a$ is given by

$$[\angle 1+\bar{\Phi}_a = (\angle Z_1+\angle Z_2+\angle Z_3)-(\angle P_1+\angle P_2+\angle P_3)] \tag{8.8}$$

(phase change for one revolution of s')

When s' is moved one complete revolution around the locus in a clockwise direction the phase change of vector $\mathbf{Z}_1$ is 0°. This is also the case for vectors $\mathbf{Z}_3$, $\mathbf{P}_1$, $\mathbf{P}_2$ and $\mathbf{P}_3$. Vector $\mathbf{Z}_2$ has a phase change of 360° for one clockwise revolution of s'. Therefore equation 8.8 becomes,

$$[\angle 1+\overline{\Phi}_a = (0°+360°+0°)-(0°+0°+0°)]$$

(phase change for one revolution of s')

$$\angle 1+\overline{\Phi}_a = 360°$$

Thus, the circular locus on the s-plane is transformed onto the $(1+\overline{\Phi}_a)$-plane, such that the locus of $1+\overline{\Phi}_a$ changes phase 360°, i.e. the locus must enclose the origin.

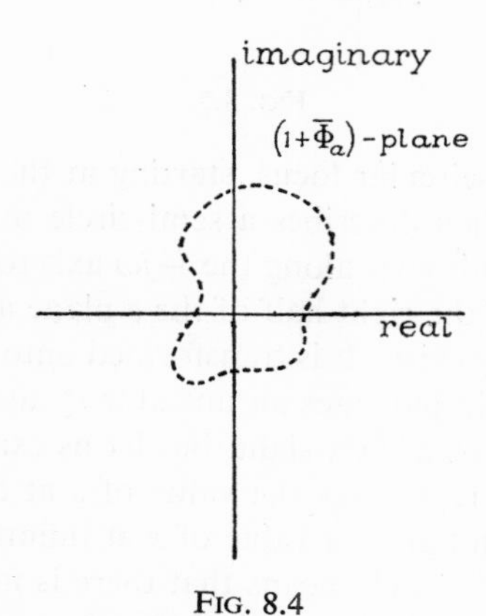

FIG. 8.4

Figure 8.4. This is a possible locus of $1+\overline{\Phi}_a$. It is obtained by transforming the circular locus, for the poles and zeros describing $1+\overline{\Phi}_a$, onto the $(1+\overline{\Phi}_a)$-plane. Equation 8.8 shows that if the locus on the s-plane does not enclose a zero (i.e. a root of $\overline{D}+K\overline{N}$) there will be no resultant 360° phase change. The locus on the $(1+\overline{\Phi}_a)$-plane will therefore not enclose the origin. We can see from this that to check whether any roots of $\overline{D}+K\overline{N}$ lie in the right half of the s-plane we enclose the whole of the right half with a locus. To indicate a stable system, transformation of this locus onto the $(1+\overline{\Phi}_a)$-plane should give a locus that does not enclose the origin. A locus enclosing the whole of the right half of the s-plane is unrealistic. We approximate to this by taking a large semi-circular locus which encloses the majority of the right half.

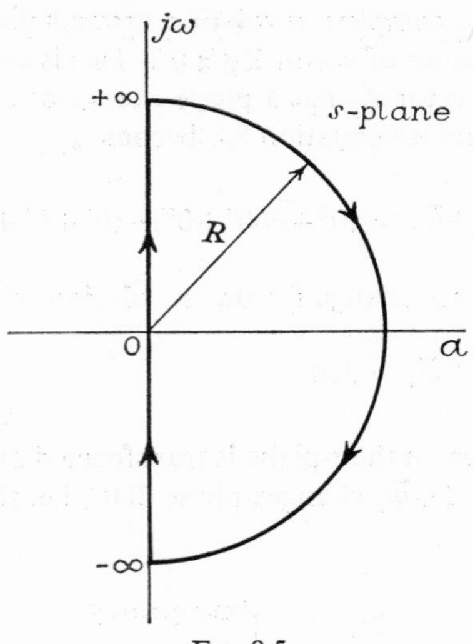

FIG. 8.5

Figure 8.5. The semi-circular locus, starting at the origin, goes along the $+j\omega$ axis to $+\infty$. It then describes a semi-circle until it reaches the $-j\omega$ axis at $-\infty$, after which it goes along the $-j\omega$ axis to the origin. This locus encloses the majority of the right half of the s-plane and is bound to enclose any realistic roots if they exist. It is transformed onto the $(1+\overline{\Phi}_a)$-plane, and encirclement of the origin indicates an unstable system. The transformation appears to be complicated at first sight, but let us examine the semi-circular locus a little closer. R represents the value of s at any point on the semi-circle's circumference and gives a value of s at infinity. Substituting R into $1+\overline{\Phi}_a$ gives $1+\overline{\Phi}_a = 1$,† which means that there is no phase change on this part of the locus. Therefore the whole 360° phase change, which a closed path locus traverses, must occur along the $j\omega$-axis from $-\infty$ to $+\infty$. Hence, to effectively enclose the whole of the right side of the s-plane, it is only necessary to consider the locus on the s-plane along the $j\omega$-axis from $-\infty$ to $+\infty$. We are now plotting the locus of $1+\overline{\Phi}_a$ on the $(1+\overline{\Phi}_a)$-plane for $s = j\omega$, where ω is $-\infty$ to $+\infty$. This is in fact the frequency response of $1+\overline{\Phi}_a$. If this locus encloses the origin the system has roots on the right side of the s-plane and is unstable. In practice it is difficult to obtain the frequency response of $1+\overline{\Phi}_a$. However we can quite easily obtain the frequency response of $\overline{\Phi}_a$ (the open-loop transfer function, which as $s = j\omega$, we may now call $\Phi_a(j\omega)$). A locus on the $\overline{\Phi}_a$-plane gives the locus on the $(1+\overline{\Phi}_a)$-plane if it is measured from the $(-1, j0)$ point on the $\overline{\Phi}_a$-plane. [The $(-1, j0)$ point is the origin of the $(1+\overline{\Phi}_a)$-plane on the $\overline{\Phi}_a$-plane.] The plot of the frequency

† The order of the numerator is less than that of the denominator ($n < m$).

response (from $\omega = -\infty$ to $\omega = +\infty$) for the open-loop transfer function as a locus in the complex plane is called a *Nyquist diagram* after H. Nyquist. The *Nyquist stability criterion* states that if the frequency response locus of the open-loop transfer function of a single-loop system encloses the $(-1, j0)$ point the system is unstable.

Various frequency response loci are shown in the following diagrams.

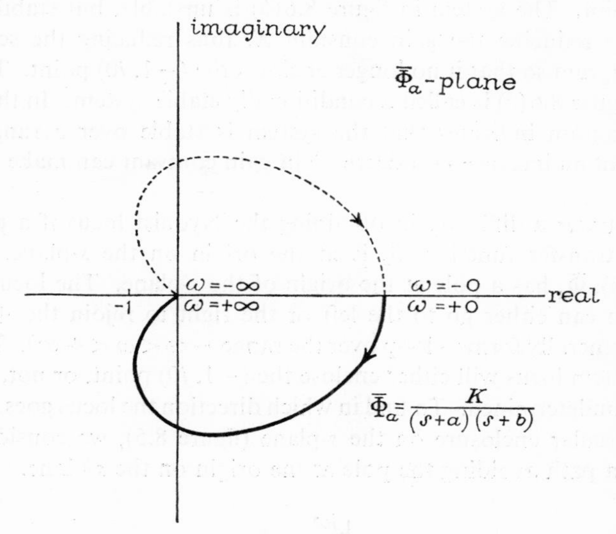

FIG. 8.6 (*a*)

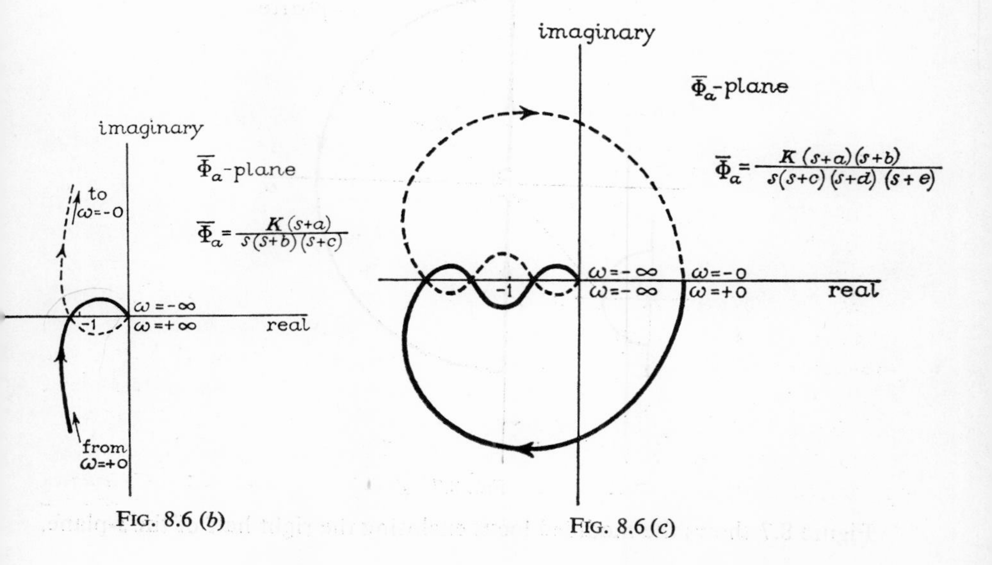

FIG. 8.6 (*b*)

FIG. 8.6 (*c*)

Figures 8.6(*a*), (*b*) and (*c*) show Nyquist diagrams for three systems. We can see that it is only necessary to plot the frequency response for frequencies 0 to $+\infty$, because the range 0 to $-\infty$ gives a mirror image of the 0 to $+\infty$ locus. However, a sketch of the complete locus, frequencies $-\infty$ to $+\infty$, gives a clearer picture of whether the $(-1, j0)$ point is enclosed or not. The system shown in figure 8.6(*a*) is stable as the locus does not enclose the $(-1, j0)$ point. The system in figure 8.6(*b*) is unstable, but stability can be obtained by reducing the gain constant K, thus reducing the scale of the Nyquist diagram so that it no longer encloses the $(-1, j0)$ point. The system shown in figure 8.6(*c*) is called a conditionally stable system. In this case the Nyquist diagram indicates that the system is stable over a range of gain constant, but an increase or a decrease in gain contsant can make the system unstable.

We encounter a difficulty in obtaining the Nyquist locus if a pole of the open-loop transfer function $\overline{\Phi}_a$ is at the origin on the s-plane. Consider figure 8.6(*b*), $\overline{\Phi}_a$ has a pole at the origin of the s-plane. The locus from the $-\omega$ branch can either go to the left or the right to rejoin the $+\omega$ branch (the locus generally forms a loop over the range $-\infty < \omega < +\infty$). This means that the system locus will either enclose the $(-1, j0)$ point, or not, and hence stability is undetermined. To find in which direction the locus goes, instead of the semi-circular enclosure on the s-plane (figure 8.5), we consider a locus which has a path avoiding the pole at the origin on the s-plane.

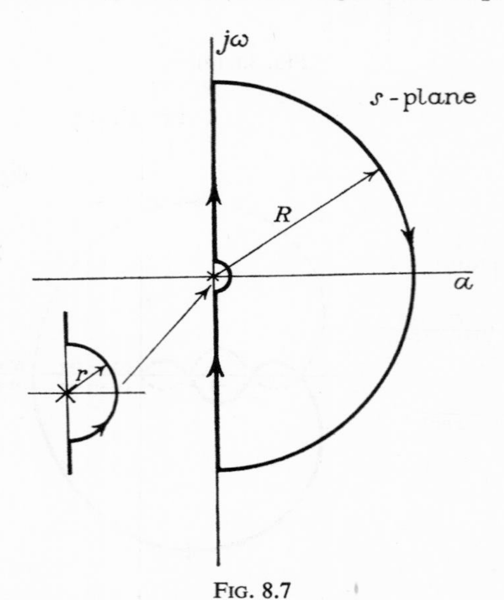

FIG. 8.7

Figure 8.7 shows the modified locus enclosing the right half of the s-plane.

The very small semi-circular part around the pole at the origin has a radius r, where r tends to zero. In polar form this semi-circle can be described by $s = r \angle\theta°$, where θ is $+90°$ to $-90°$.

Consider the open-loop transfer function $\overline{\Phi}_a$ given in figure 8.6(*b*).

$$\overline{\Phi}_a = \frac{K(s+a)}{s(s+b)(s+c)}$$

as r tends to zero

$$\overline{\Phi}_a = \frac{Ka}{sbc}$$

$$\overline{\Phi}_a = \frac{K'}{s}$$

where Ka/bc is the constant K'. Since $s = r \angle\theta°$,

$$\overline{\Phi}_a = K'/r \angle -\theta° \tag{8.9}$$

Equation 8.9 is the equation for the small semi-circular locus transformed onto the $\overline{\Phi}_a$-plane. It gives the effect of a pole at the origin. ω tends to zero therefore $s = j\omega$ tends to zero and K'/r tends to infinity. This gives the locus of $\overline{\Phi}_a$ as a semi-circle of a large radius going from $+90°$ to $-90°$ with respect to the real axis.

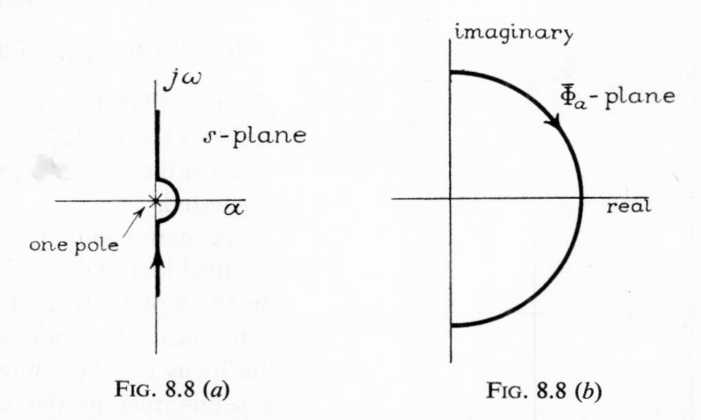

FIG. 8.8 (*a*) FIG. 8.8 (*b*)

Figures 8.8(*a*) and (*b*) show, respectively, the semi-circular locus on the *s*-plane and its transformation onto the $\overline{\Phi}_a$-plane when $\overline{\Phi}_a$ has one pole at the origin of the *s*-plane.

It can be shown, similarly, that if an open-loop transfer function has two poles at the origin of the *s*-plane, the transformation of the small semi-circular locus onto the $\overline{\Phi}_a$-plane gives a 360° clockwise circular locus of large radius.

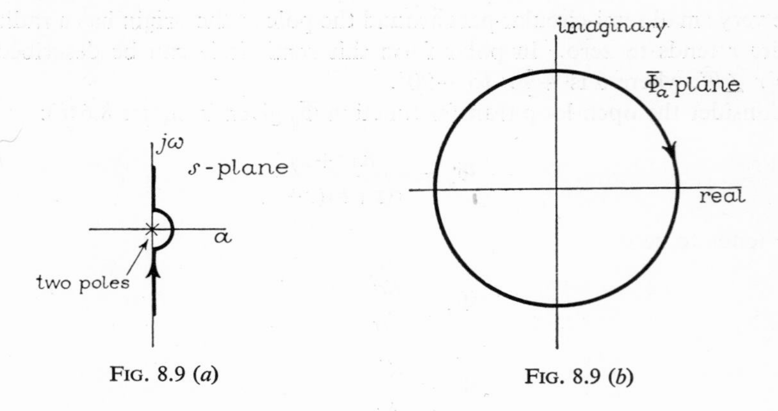

FIG. 8.9 (a) FIG. 8.9 (b)

Figures 8.9(*a*) and (*b*) show, respectively, the semi-circular locus on the *s*-plane and its transformation onto the $\bar{\Phi}_a$-plane when $\bar{\Phi}_a$ has two poles at the origin of the *s*-plane. (Each further pole at the origin of the *s*-plane provides a further semi-circular clockwise rotation on the $\bar{\Phi}_a$-plane.)

The locus shown in figure 8.6(*b*) is completed by a semi-circle of infinite radius, which starts from the $-\omega$ branch and goes around the origin in a clockwise direction. In practice it may be completed with a large semi-circle.

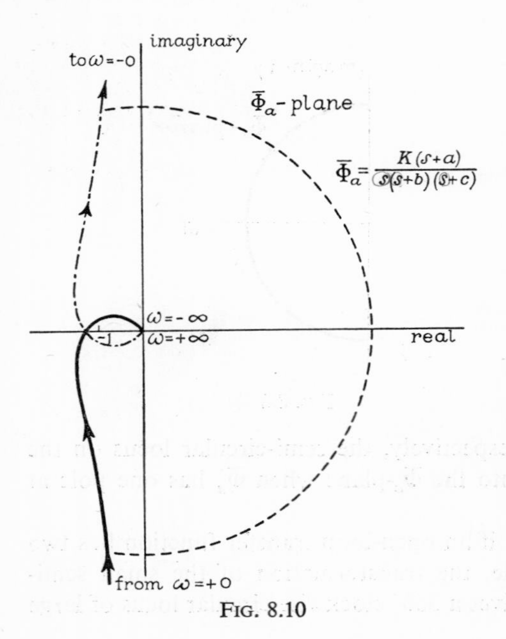

FIG. 8.10

Figure 8.10 shows the complete locus for the system

$$\bar{\Phi}_a = K(s+a)/[s(s+b)(s+c)]$$

(figure 8.6(*b*)). The system encloses the $(-1, j0)$ point and is unstable as was previously suggested.

We can conclude that if a Nyquist locus with a single pole on the *s*-plane tends to a large value near the imaginary axis, the locus can be completed by a semi-circle in the clockwise direction from the $-\omega$ branch. Similarly, if a Nyquist locus with a double pole on the *s*-plane tends to a large value near the real axis, the locus can be completed by a circle in a clockwise direction from the branch.

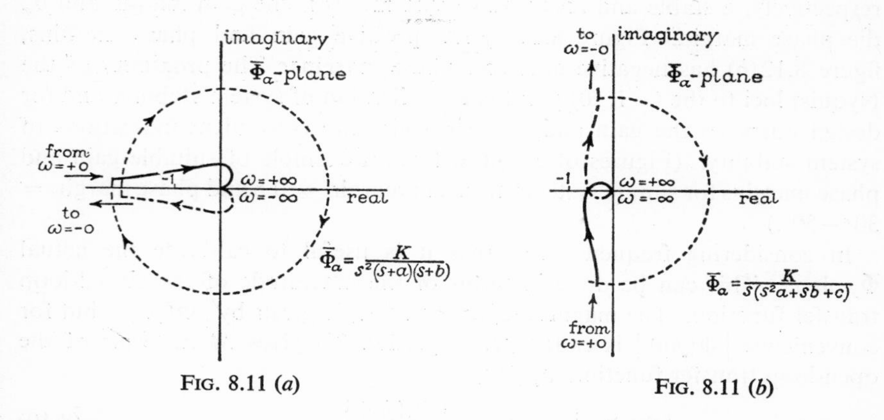

FIG. 8.11 (*a*) FIG. 8.11 (*b*)

Figures 8.11 (*a*) and (*b*) illustrate the two conclusions for the completion of Nyquist loci with pole(s) at the origin of the *s*-plane. Both systems can be unstable or stable according to the value of the gain constant *K*.

8.3 Calibration of the Nyquist diagram

Two guides to the stability of a system are the *gain margin* and the *phase margin* on Nyquist diagrams. A circle, centre at the origin and unity radius, is drawn on the plane. The gain margin is the distance, along the negative real axis, between the intersection of this axis by the locus and the $(-1, j0)$ point. The phase margin is the angle between the negative real axis and the intersection of the locus with the unity radius circle.

Figures 8.12 (*a*) and (*b*) show part of the Nyquist loci, which represent,

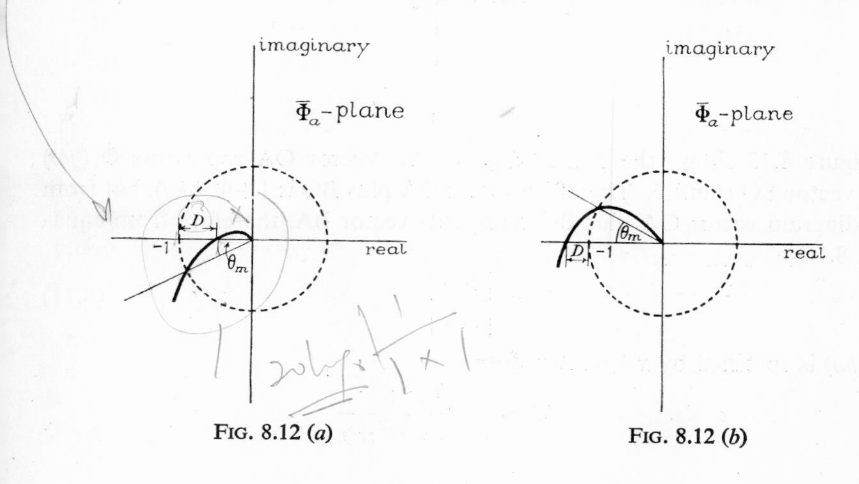

FIG. 8.12 (*a*) FIG. 8.12 (*b*)

respectively, a stable and an unstable system. D is the gain margin and θ_m the phase margin. Figure 8.12(*a*) has positive gain and phase margins, figure 8.12(*b*) has negative gain and phase margins. The proximity of the Nyquist loci to the $(-1, j0)$ point is an indication of system stability and for design purposes the gain and phase margins are convenient indications of system stability. (Figures often quoted as an example of suitable gain and phase margins for simple systems are: gain margin = 0·6 and phase margin = 30°–50°.)

In considering frequency response it is useful to calibrate the actual $\overline{\Phi}_a$-plane. This can be done in terms of the magnitude of the closed-loop transfer function. The magnitude, or modulus, is given by $|\Phi(j\omega)|$, but for convenience $|\Phi(j\omega)|$ is called M. Equation 8.5 gives M in terms of the open-loop transfer function $\Phi_a(j\omega)$,

$$|\Phi(j\omega)| = M = \frac{|\Phi_a(j\omega)|}{|1+\Phi_a(j\omega)|} \tag{8.10}$$

In order to calibrate the Φ_a-plane we must find an expression for the loci of different values of constant M, i.e. M contours. In general, on the $\overline{\Phi}_a$-plane, any point on the $\Phi_a(j\omega)$ locus can be specified in the complex form $x+jy$. Let us consider any point (x, jy) on a Nyquist locus.

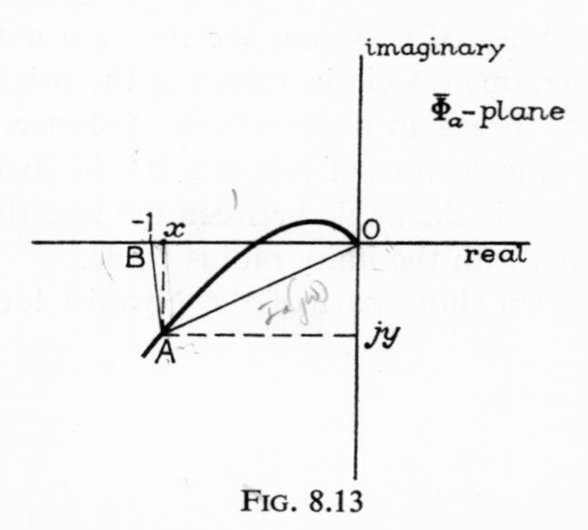

FIG. 8.13

Figure 8.13 shows the point (x, jy) at A. Vector OA represents $\Phi_a(j\omega)$ and vector BO is unity. Therefore vector OA plus BO is $1+\Phi_a(j\omega)$, but from the diagram vector OA plus BO represents vector BA, therefore from equation 8.10

$$M = \frac{|\mathrm{BO}|}{|\mathrm{BA}|} \tag{8.11}$$

$\Phi_a(j\omega)$ is specified by $x+jy$, therefore

$$M = \frac{\sqrt{(x^2+y^2)}}{\sqrt{[(1+x)^2+y^2]}}$$

$$y^2+x^2+(2x+1)\frac{M^2}{M^2-1} = 0 \tag{8.12}$$

Equation 8.12, for a particular value of M, is a circle (except for $M = 1$, when it is a straight line). The centre of the circle is at

$$y = 0$$

$$x = -M^2/(M^2-1)$$

and its radius is $M/(M^2-1)$. The M contours calibrate the $\overline{\Phi}_a$-plane as circles.

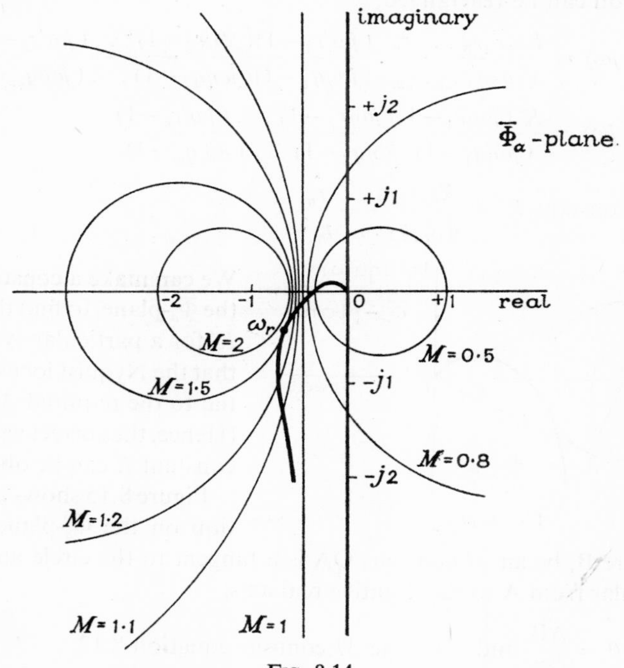

FIG. 8.14

Figure 8.14. M contours are drawn directly on the $\overline{\Phi}_a$-plane. A particularly useful value of magnitude M, as we shall see later, is when the Nyquist locus is tangential to an M contour. This gives the system's maximum magnitude, M_{max}, and resonant frequency, i.e. the maximum magnitude occurs at the particular frequency ω_r. The Nyquist locus in figure 8.14 shows the M_{max} contour at $M = 1{\cdot}5$.

We can draw phase calibrations on the $\overline{\Phi}_a$-plane in a similar way to magnitude calibrations. Phase contours are circles called N contours. We shall not consider them in detail as they are not necessary in our discussion.

It is often convenient to use the inverse Nyquist diagram. This is a plot of the locus of $1/\Phi_a(j\omega)$.

Maximum magnitude M_{max} and its relationship with the gain constant K

The value of M_{max} for a particular system gives a guide to the system performance. If a particular M_{max} value is specified for a system it can only be obtained if the gain constant K is set correctly. First let us consider a general open-loop transfer function

$$\Phi_a(j\omega) = \frac{K(j\omega-r_1)(j\omega-r_2)(j\omega-r_3)\ldots(j\omega-r_n)}{(j\omega-q_1)(j\omega-q_2)(j\omega-q_3)\ldots(j\omega-q_m)}$$

This equation can be rearranged,

$$\Phi_a(j\omega) = \frac{Kr_1r_2r_3\ldots r_n}{q_1q_2q_3\ldots q_m}\cdot\frac{(j\omega/r_1-1)(j\omega/r_2-1)\ldots(j\omega/r_n-1)}{(j\omega/q_1-1)(j\omega/q_2-1)\ldots(j\omega/q_m-1)}$$

$$\Phi_a(j\omega) = \frac{K'(j\omega/r_1-1)(j\omega/r_2-1)\ldots(j\omega/r_n-1)}{(j\omega/q_1-1)(j\omega/q_2-1)\ldots(j\omega/q_m-1)}$$

where the constant $K' = \dfrac{Kr_1r_2r_3\ldots r_n}{q_1q_2q_3\ldots q_m}$

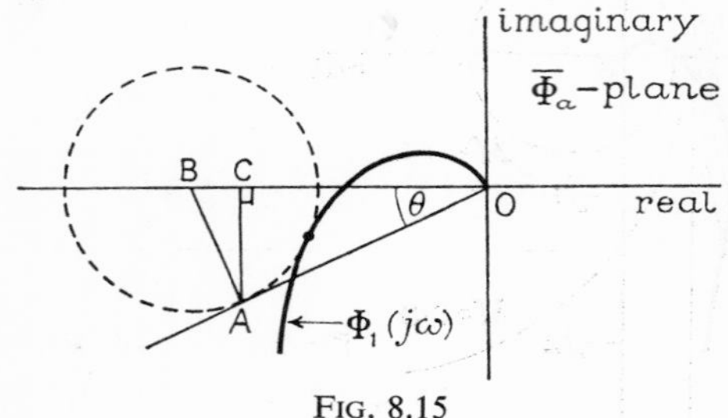

FIG. 8.15

We can make a construction on the $\overline{\Phi}_a$-plane, to find the value of K' for a particular system, such that the Nyquist locus is tangential to the required M contour. (Hence, the correct value of gain constant K can be obtained.)

Figure 8.15 shows a construction on the $\overline{\Phi}_a$-plane. Let the circle, centre B, be an M contour, OA is a tangent to the circle and AC is a perpendicular from A to the negative real axis.

$$\sin\theta = \frac{\text{AB}}{\text{OB}}$$ and from the M contour equation 8.12,

$$\text{AB} = \frac{M}{M^2-1} \quad \text{and} \quad \text{OB} = \frac{-M^2}{M^2-1}, \text{ therefore,}$$

$$\sin\theta = \frac{1}{M} \tag{8.13}$$

By Pythagoras, $$\text{AO} = \sqrt{(\text{OB}^2-\text{AB}^2)}$$

therefore $$\text{AO} = \frac{M}{\sqrt{(M^2-1)}}$$

and $$\text{CA} = \frac{M}{\sqrt{(M^2-1)}}\sin\theta$$

From equation 8.13 $\quad \text{CA} = \dfrac{1}{\sqrt{(M^2-1)}}$

$$\text{OC} = \sqrt{(\text{AO}^2-\text{CA}^2)}$$

therefore $\quad \text{OC} = \sqrt{\left(\dfrac{M^2}{M^2-1}-\dfrac{1}{M^2-1}\right)}$

$$\text{OC} = 1$$

therefore, C is the $(-1, j0)$ point.

We shall now consider the construction in conjunction with a Nyquist diagram. Suppose that a particular value of M_{max} is required for a system whose open-loop transfer function is $\Phi_a(j\omega) = K'\Phi'(j\omega)$. The constant K' which gives the required M_{max} must be determined. First K' is adjusted to unity and the Nyquist locus of $\Phi'(j\omega)$ plotted.

Figure 8.16. Using equation 8.13 ($\theta = \sin^{-1} 1/M_{max}$) find the value of θ for the required value of M_{max} and draw line OA. By trial and error construct a circle, centre B, on the negative real axis, such that it is tangential to the $\overline{\Phi}'(j\omega)$ locus and OA. If K' is unity at the required M_{max}, OC will be unity and the circle will have a radius of $M_{max}/(M^2{}_{max}-1)$. However, it would be a coincidence if $K'=1$ were the correct constant. For the correct construction, OC must be unity. To obtain this the diagram must be scaled by a factor of 1/OC. The new Nyquist locus is (1/OC) $\overline{\Phi}'(j\omega)$ and the new circle will be the M contour of M_{max}. Thus the correct constant for a specific M_{max} is 1/OC. (The Nyquist diagram need not necessarily be drawn for $K'=1$, but, in finding the correct constant, the adjustment for length OC must be taken into consideration.)

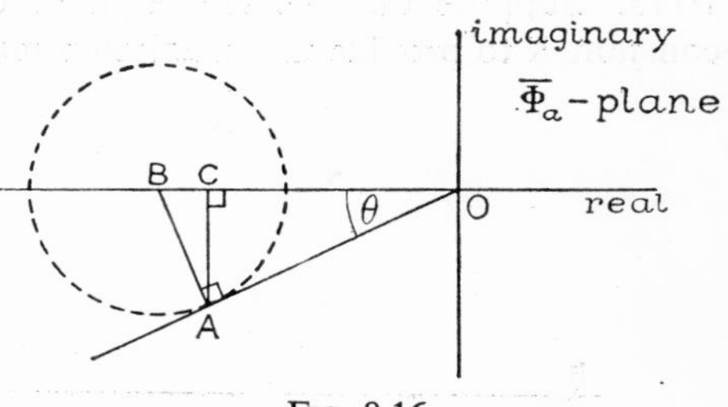

FIG. 8.16

8.4 Example

In chapter 7 we discussed the remote position control system root-locus pattern. We considered the adjustment of gain constant and the possibility of system improvement by compensation. The Nyquist diagram can be used in much the same way. By observation we know that the closer the locus comes to the $(-1, j0)$ point, without enclosing it, the smaller the damping ratio ζ becomes. In general, the phase margin gives a good indication of damping for simple systems. Specified values of gain margin and phase margin are known to provide certain system performances and the system adjusted to meet the specifications. The maximum magnitude gives a guide to the transient response and for the second-order systems can be quite easily calibrated against the damping ratio ζ. (This is considered in detail in chapter 10.)

We shall consider a system whose open-loop transfer function is

$$\bar{\Phi}_a = \frac{K}{s(s+2)(s+3)}$$

In terms of frequency

$$\Phi_a(j\omega) = \frac{K}{j\omega(j\omega+2)(j\omega+3)} = \frac{K/6}{j\omega(j\omega/2+1)(j\omega/3+1)}$$

The specification for the maximum magnitude for this system is $M_{max} = 1{\cdot}113$. Suppose that we are required to find the necessary value of gain constant K to provide this maximum magnitude.

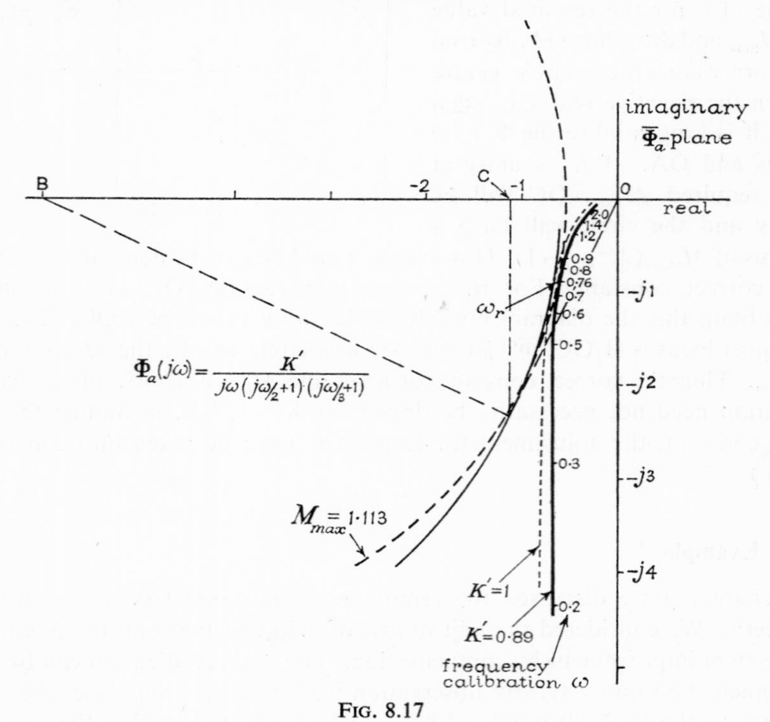

FIG. 8.17

Figure 8.17 shows the Nyquist locus of $\Phi_a(j\omega)$ for $K/6 = K' = 1$. The construction to obtain the constant K', using the relationship of equation

8.13 ($\sin^{-1} 1/M_{max} = 64°$), is shown on the complex plane. 1/OC is found to be 0·89 and hence the required gain constant is $K = 0{\cdot}89 \times 6 = 5{\cdot}34$. The required system Nyquist locus $\Phi_a(j\omega)$ is shown on the figure. The gain margin and phase margin are 0·81 and 55° respectively.

In chapter 7 we discussed briefly the use of *passive lead compensation* to make a system more stable over a range of gain constant K. The passive lead network, figure 7.22, is described by the transfer function

$$\frac{\bar{e}_o}{\bar{e}_i} = \frac{s+1/T}{s+1/TA}$$

and in terms of frequency

$$\frac{e_o}{e_i} = \frac{j\omega + 1/T}{j\omega + 1/TA}$$

In the example if the gain constant K is large the system is unstable.

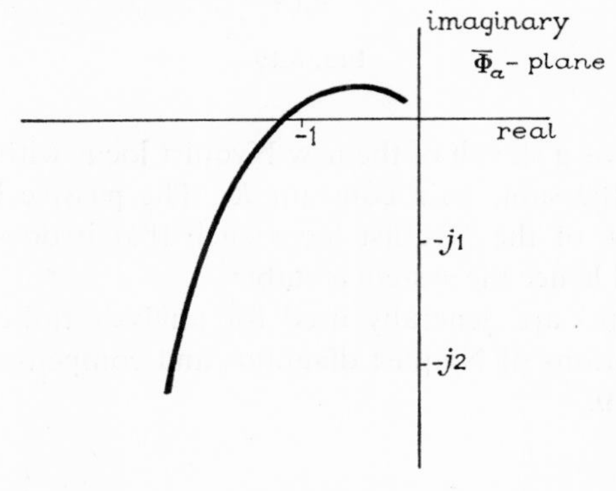

FIG. 8.18

Figure 8.18 shows a sketch of the Nyquist locus of

$$\Phi_a(j\omega) = \frac{K}{j\omega(j\omega+2)(j\omega+3)}$$

for a large value of K. When, as before, the passive lead network is inserted in the forward loop the open-loop transfer function becomes

$$\Phi_a(j\omega) = \frac{K(j\omega+1/T)}{j\omega(j\omega+2)(j\omega+3)(j\omega+1/TA)}$$

and a new Nyquist locus is obtained.

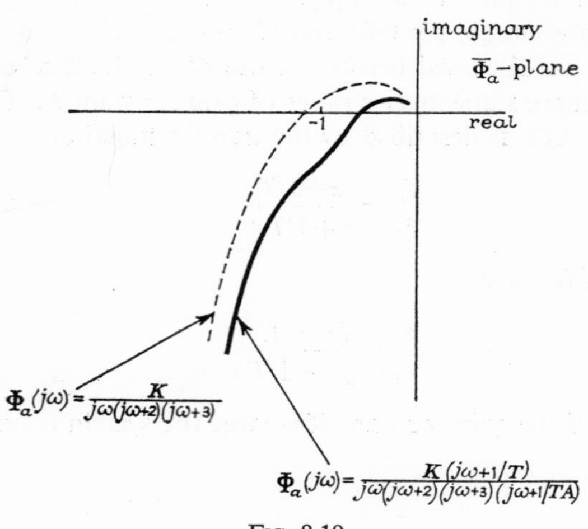

FIG. 8.19

Figure 8.19 shows a sketch of the new Nyquist locus with particular values of T and A and the same gain constant K. The passive lead network has changed the shape of the Nyquist locus such that it does not enclose the $(-1, j0)$ point and hence the system is stable.

Nyquist diagrams are generally used for analysis rather than synthesis. Further considerations of Nyquist diagrams and compensation methods are given in chapter 10.

8.5 Bode diagrams

The Nyquist diagram representation of the open-loop frequency response of a control system does not, itself, readily lead to system design. Other means of showing frequency response can help with the overall information to be obtained about a system.

The Nyquist locus is the plot of three quantities: magnitude, phase and frequency. It can be split into two diagrams, one the graph of magnitude against frequency, the other the graph of phase against frequency. (It is important to remember that magnitude refers to the output of the open-loop for a 'unity' r.m.s. sinusoidal input, i.e. magnitude is $|\Phi_a(j\omega)|$.) The two graphs used are the graph of the logarithm of the magnitude against the logarithm of the frequency and the graph of phase against the logarithm of the frequency.

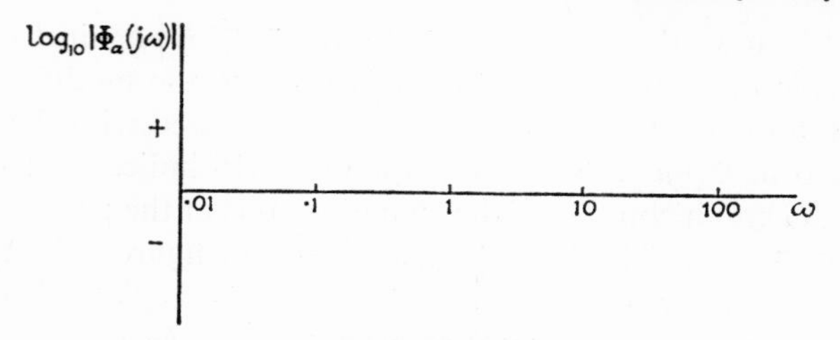

FIG. 8.20 (*a*)

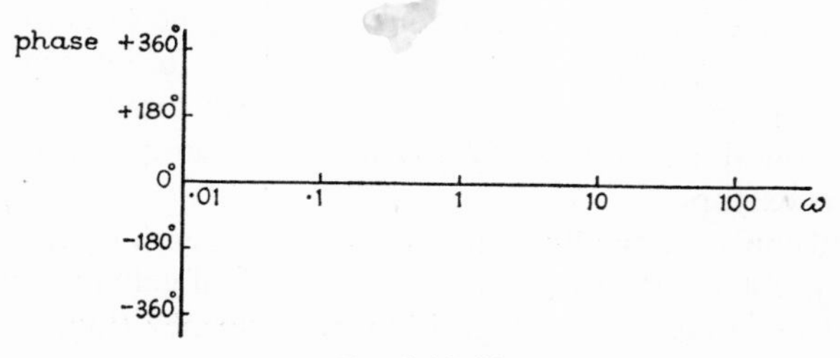

FIG. 8.20 (*b*)

Figures 8.20(*a*) and (*b*) show the axes of the two graphs used to show system frequency response. The graphs are called *Bode diagrams* after H. W. Bode.

The reason for the use of logarithms becomes evident if we consider a specific example of an open-loop transfer function $\Phi_a(j\omega)$,

$$\Phi_a(j\omega) = \frac{(j\omega T_1+1)}{(j\omega T_2+1)(j\omega T_3+1)} \tag{8.14}$$

The example, equation 8.14, can be expressed as two equations, the modulus of the open-loop transfer function

$$|\Phi_a(j\omega)| = \frac{\sqrt{(\omega^2 T_1^2+1)}}{[\sqrt{(\omega^2 T_2^2+1)}]\,[\sqrt{(\omega^2 T_3^2+1)}]} \tag{8.15}$$

and the phase of the open-loop transfer function,

$$\theta = \frac{\angle \tan^{-1} \omega T_1}{\angle \tan^{-1} \omega T_2 \ \angle \tan^{-1} \omega T_3}$$

$$\theta = \angle \tan^{-1} \omega T_1 - (\angle \tan^{-1} \omega T_2 + \angle \tan^{-1} \omega T_3) \tag{8.16}$$

Equation 8.15 can be expressed in logarithms to the base ten as follows:

$$\log_{10} |\Phi_a(j\omega)| = \tfrac{1}{2} \log_{10} (\omega^2 T_1^2+1) - \tfrac{1}{2} \log_{10} (\omega^2 T_2^2+1) - \tfrac{1}{2} \log_{10} (\omega^2 T_3^2+1) \tag{8.17}$$

Equation 8.17 shows that the graph of $\log_{10} |\Phi_a(j\omega)|$ against $\log_{10} \omega$ consists of the sum of the graphs of the logarithms of the moduli of the poles and zeros of $\Phi_a(j\omega)$. Hence, the graph is easily altered with the addition of a pole or zero to $\Phi_a(j\omega)$. Similarly, equation 8.16 indicates that the phase against $\log_{10} \omega$ graph consists of the sum of graphs of the phases of the poles and zeros of $\Phi_a(j\omega)$. The $\log_{10} |\Phi_a(j\omega)|$-axis in figure 8.20(*a*) is usually scaled by 20. The scale factor of 20 gives $20 \log_{10} |\Phi_a(j\omega)|$, which is defined as the magnitude (or gain) measurement in *decibels* (dB). (Gain in dB $= 20 \log_{10} |\Phi_a(j\omega)|$.) A transfer function consists of poles and zeros of the form $1/T(j\omega T+1)$, where T is a constant. (A pole or zero at the origin is when $1/T = 0$.) It follows that if we know how to construct the graphs for a pole and a zero of the form $(j\omega T+1)$ we can draw the graphs of any transfer function simply by adding the individual graphs together. Let us consider some examples.

(*i*) We shall construct the Bode diagrams of the open-loop transfer function $\Phi_a(j\omega) = 1/(j\omega T+1)$. We shall consider the graph of gain in decibels (gain $= 20 \log_{10} |\Phi_a(j\omega)|$) against the logarithm of frequency ($\log_{10} \omega$) first. (For convenience this graph will be referred to as the *gain graph*.) The gain in decibels, at any frequency ω, is given by

$$\text{gain} = 20 [-\log_{10}\sqrt{(\omega^2T^2+1)}]$$

$$\text{gain} = -10 \log_{10} (\omega^2T^2+1) \tag{8.18}$$

Equation 8.18 gives the gain graph, but the graph can be usefully approximated by asymptotes. Consider the range of frequency when ω is very much less than unity. Equation 8.18 becomes

$$[\text{gain}]_{\omega \ll 1} = -10 \log_{10} 1$$

therefore,

$$[\text{gain}]_{\omega \ll 1} = 0 \tag{8.19}$$

Now, let ω be very much larger than unity. Equation 8.18 becomes

$$[\text{gain}]_{\omega \gg 1} = -10 \log_{10} \omega^2T^2$$

$$[\text{gain}]_{\omega \gg 1} = -20 \log_{10} \omega T \tag{8.20}$$

Equations 8.19 and 8.20 represent two straight lines making up an approximation to the gain graph for $\Phi_a(j\omega) = 1/(j\omega T+1)$. Equation 8.19 represents a straight line along the $\log_{10} \omega$ axis terminating at the start of equation 8.20. Equation 8.20 can be rewritten

$$[\text{gain}]_{\omega \gg 1} = -20 \log_{10} T - 20 \log_{10} \omega$$

which is the equation of a straight line of slope -20 intersecting the $\log_{10} \omega$ axis at $\omega = 1/T$. It should be noted that as the base of logarithm is 10, the calibration of the $\log_{10} \omega$ axis is in decades and the slope is, in fact, -20dB per decade (this is also a slope of -6dB per octave).

Figure 8.21 shows the gain graph approximated by two asymptotes (T has been chosen as unity). From equations 8.19 and 8.20 the greatest inaccuracy of the approximations will occur at the meeting point of the two asymptotes. This point is called the *corner point*. The error at the corner frequency is found by substituting $\omega = 1/T$ into equation 8.18, therefore

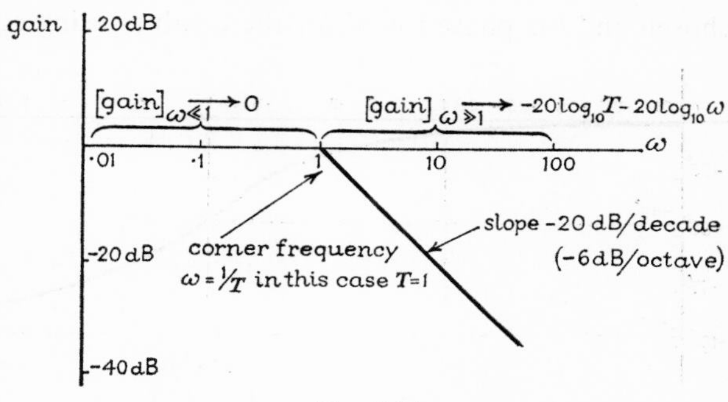

FIG. 8.21

$$[\text{gain}]_{\omega = 1/T} = -10 \log_{10} (2)$$

$$[\text{gain}]_{\omega = 1/T} \simeq -3\text{dB}$$

This means that the corner point is really 3dB below its approximate position. Figure 8.22. A more accurate graph can now be drawn, as shown in the figure. The gain graph obtained by the asymptotes and the deviation at the corner point usually provide a useful approximation.

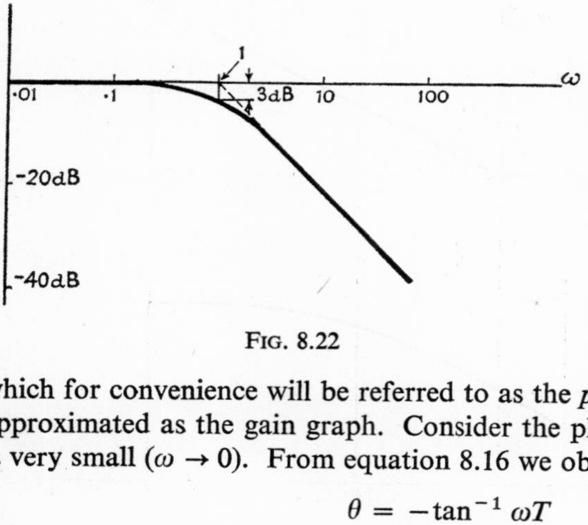

FIG. 8.22

The graph of the phase against the logarithm of the frequency ($\log_{10} \omega$), which for convenience will be referred to as the *phase graph*, is not as easily approximated as the gain graph. Consider the phase θ when the frequency is very small ($\omega \to 0$). From equation 8.16 we obtain

$$\theta = -\tan^{-1} \omega T \tag{8.21}$$

$$[\theta]_{\omega \to 0} = 0$$

Now, when the frequency is very large ($\omega \to \infty$) we obtain

$$[\theta]_{\omega \to \infty} = -\tan^{-1} \infty = -90^\circ$$

The two values of phase obtained show us that for small frequencies the phase tends to 0° and for large frequencies it tends to −90°. To complete a reasonable phase graph values of frequency over a useful range must be chosen and the phase found for these values using equation 8.21.

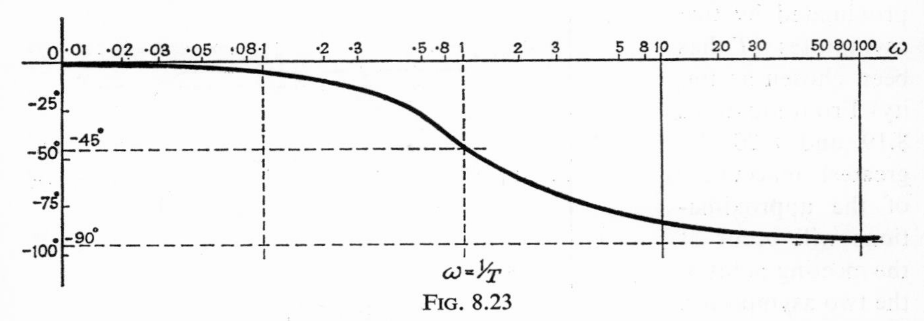

FIG. 8.23

Figure 8.23. The figure is the phase graph. It shows a −45° phase angle when $\omega = 1/T$ (T has been chosen as unity as an example).

Bode diagrams consist of the pair of graphs, the gain and phase graphs.

(*ii*) The Bode diagrams of the open-loop transfer function $\Phi_a(j\omega) = (1+j\omega T)$ are constructed in the same way as for example (*i*). However, the slope of the asymptote after the corner point is positive, and the phase angle is always positive.

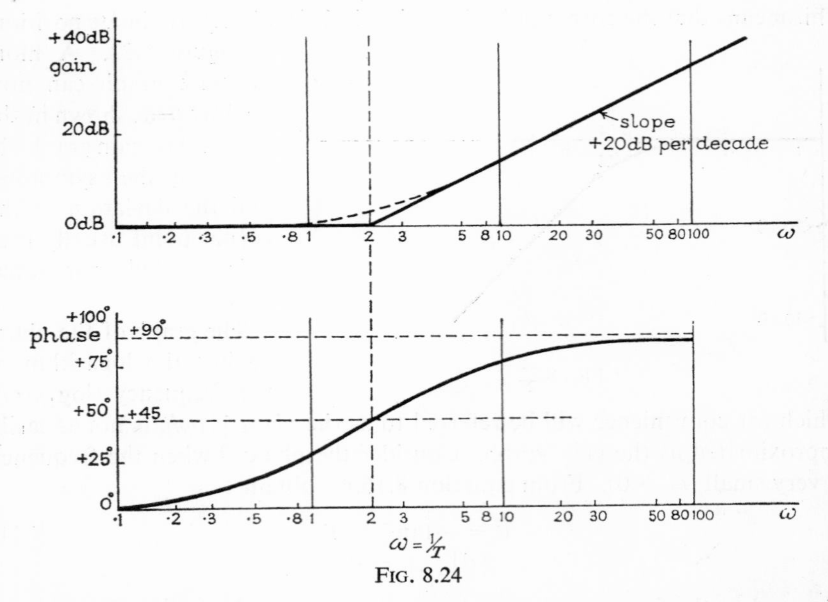

FIG. 8.24

Figure 8.24. The figure shows the Bode diagrams for $\Phi_a(j\omega) = (1+j\omega T)$ (T has been chosen as 0·5 as an example).

(*iii*) The sum of the gain graphs and the sum of the phase graphs of figures 8.21, 8.23 and 8.24 gives the Bode diagrams of the transfer function

$$\Phi_a(j\omega) = \frac{(1+j\omega 0{\cdot}5)}{(1+j\omega)}$$

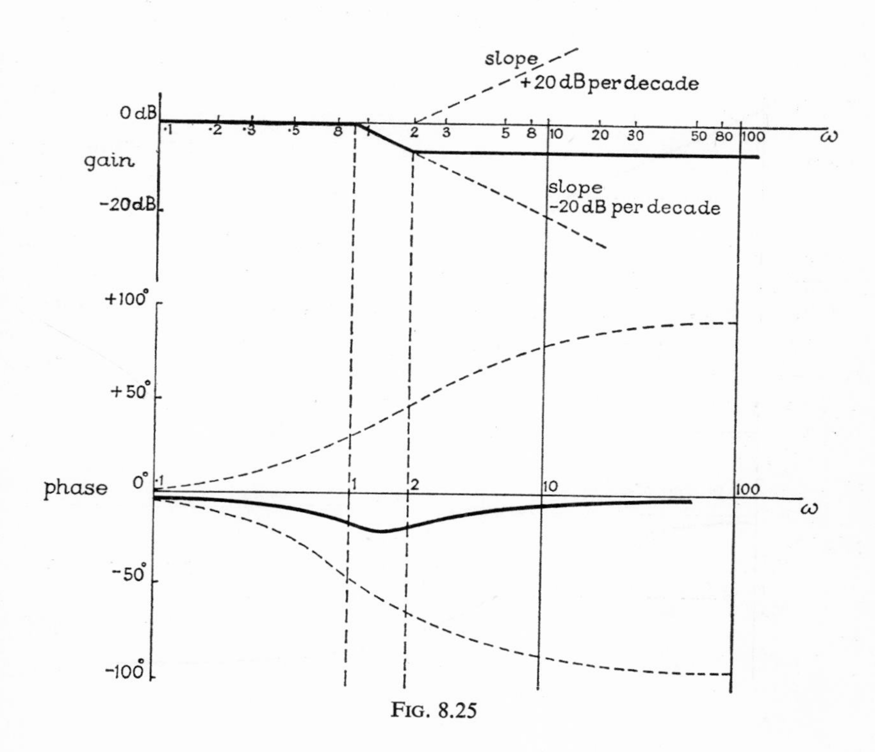

FIG. 8.25

Figure 8.25 shows the Bode diagrams for the transfer function $\Phi_a(j\omega) = (1+j\omega 0{\cdot}5)/(1+j\omega)$. The individual slopes of -20dB per decade and $+20$dB per decade of the pole and zero cancel after the second corner point at $\omega = 1/0{\cdot}5 = 2$.

8.6 Calibration of the Bode diagrams

Calibration is best approached by considering a specific example. We shall examine the Bode diagrams of the system whose open-loop transfer function is

$$\Phi_a(j\omega) = \frac{K}{j\omega(j\omega+1)(j\omega+4)}$$

This transfer function can be rearranged as follows

$$\Phi_a(j\omega) = \frac{K/4}{j\omega(j\omega+1)(j\omega 0{\cdot}25+1)}$$

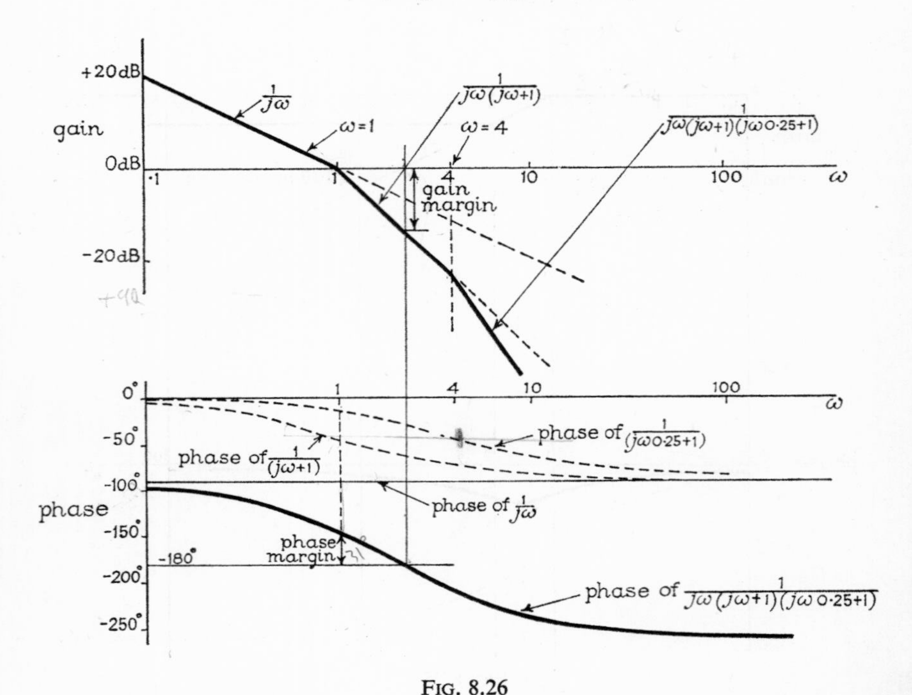

FIG. 8.26

Figure 8.26 shows the Bode diagrams for the open-loop transfer function

$$\Phi_a(j\omega) = \frac{K'}{j\omega(j\omega+1)(j\omega 0{\cdot}25+1)}$$

The diagrams have been approximated for a constant $K' = 1$. The diagrams are not calibrated in terms of the gain constant, because a change in gain is easily effected by raising the gain graph above the $\log_{10} \omega$ axis by the appropriate amount of $20 \log_{10} K'$. The gain margin is the modulus of $\Phi_a(j\omega)$, when the phase is 180°, minus unity. On the gain graph the gain margin, in decibels, is the value of gain at the frequency for which the phase is 180°.

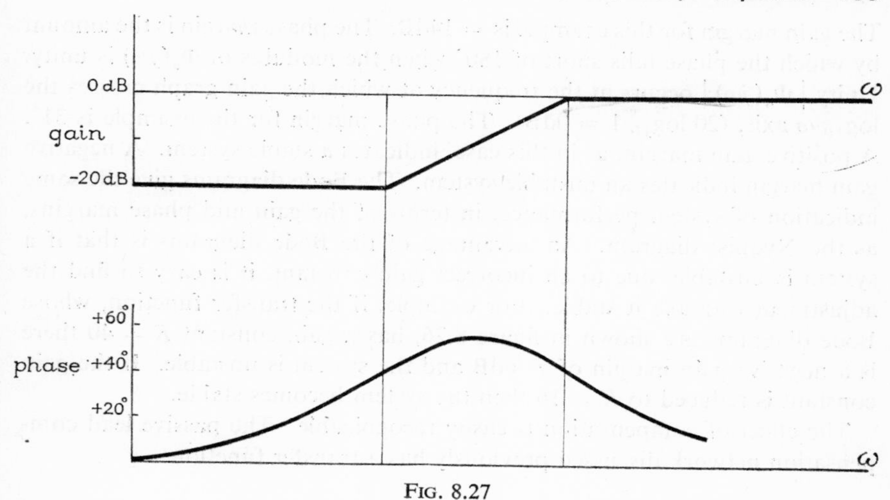

FIG. 8.27

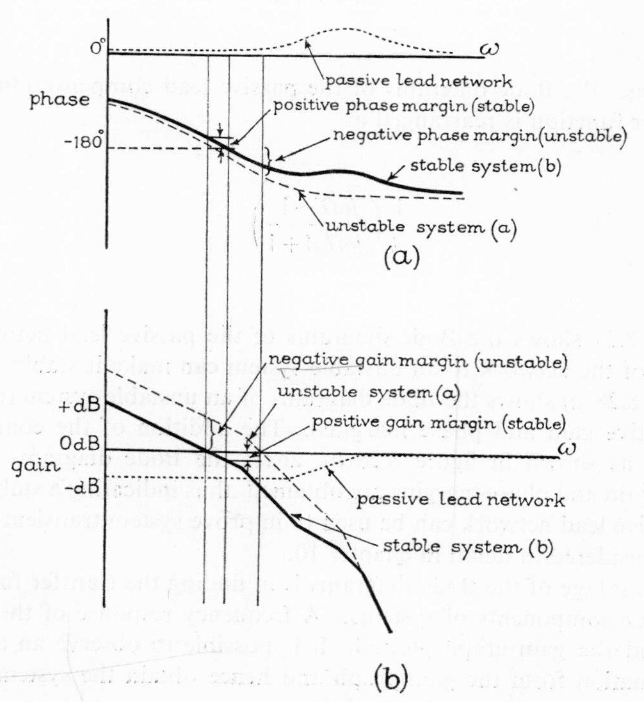

FIG. 8.28

The gain margin for this example is $\doteq$ 14dB. The phase margin is the amount by which the phase falls short of 180° when the modulus of $\Phi_a(j\omega)$ is unity. Unity $|\Phi_a(j\omega)|$ occurs at the frequency at which the gain graph crosses the $\log_{10} \omega$ axis, ($20 \log_{10} 1 = 0$dB). The phase margin for the example is 31°. A positive gain margin, as in this case, indicates a stable system. A negative gain margin indicates an unstable system. The Bode diagrams give the same indication of system performance, in terms of the gain and phase margins, as the Nyquist diagram. An advantage of the Bode diagrams is that if a system is unstable, due to an incorrect gain constant, it is easy to find the adjustment to make it stable. For example, if the transfer function, whose Bode diagrams are shown in figure 8.26, has a gain constant $K = 40$ there is a negative gain margin of $\doteq$ 6dB and the system is unstable. If the gain constant is reduced to $K = 16$ then the system becomes stable.

The effect of compensation is easily recognizable. The passive lead compensation network discussed previously has a transfer function

$$\frac{j\omega + 1/T}{j\omega + 1/TA}$$

To construct the Bode diagrams of the passive lead compensation network its transfer function is rearranged as

$$\frac{1}{A}\left(\frac{j\omega T + 1}{j\omega TA + 1}\right)$$

Figure 8.27 shows the Bode diagrams of the passive lead network. The addition of the network to an unstable system can make it stable.

Figure 8.28(*a*) shows the Bode diagrams of an unstable system (the system has negative gain and phase margins). The addition of the compensation network, as shown in figure 8.28(*b*), alters the Bode diagrams such that positive gain and phase margins are obtained, thus indicating a stable system. The passive lead network can be used to improve system transient response; this is considered in detail in chapter 10.

An advantage of the Bode diagrams is in finding the transfer functions of systems or components of systems. A frequency response of the system is taken and the gain graph plotted. It is possible to observe an asymptotic approximation from the gain graph and hence obtain the system's transfer function.

8.7 Nichols diagrams

Another graphical representation of the open-loop frequency response, which is useful in obtaining information about a system, is the Nichols diagram. This is a plot of the gain in decibels against the phase in degrees. To illustrate the Nichols diagram we shall use the example of the third-order system of the previous section.

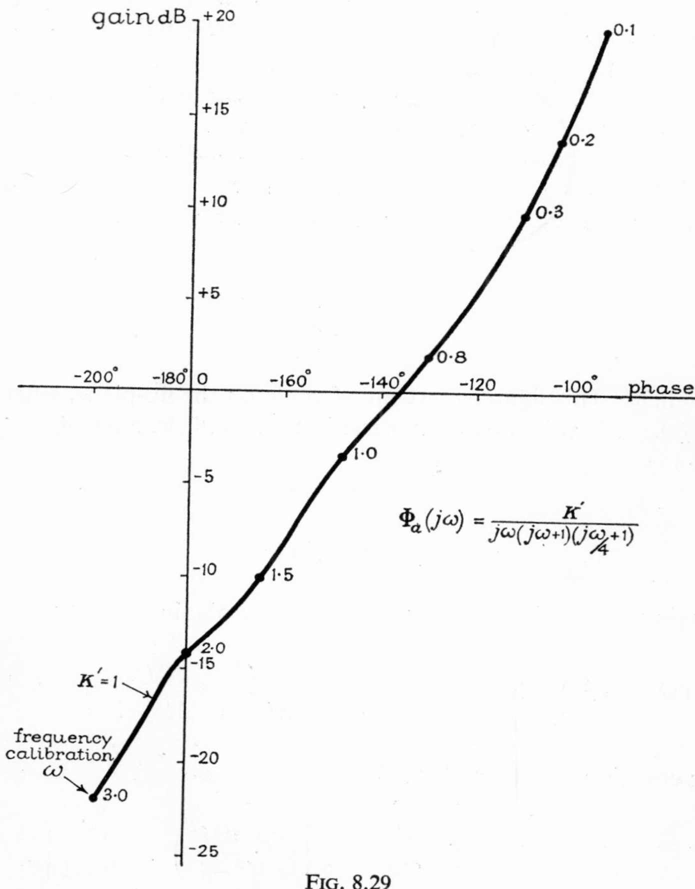

FIG. 8.29

Figure 8.29 shows the Nichols diagram for the open-loop transfer function

$$\Phi_a(j\omega) = \frac{K'}{j\omega(j\omega+1)(j\omega 0{\cdot}25+1)}$$

with the constant $K' = 1$.

The Nichols diagram is calibrated in the same way as the Nyquist diagram. Constant magnitude contours are drawn on the Nichols diagram. The equation for the M circles on the Nyquist diagram at a point (x, jy) is

$$y^2+x^2+(2x+1)\frac{M^2}{M^2-1} = 0 \tag{8.12}$$

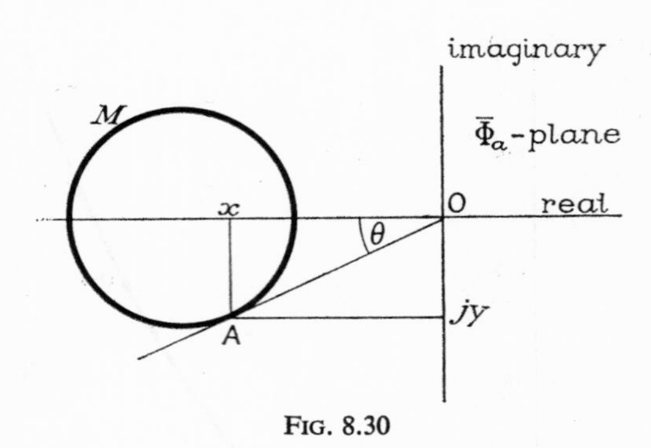

FIG. 8.30

Figure 8.30. This figure shows an M circle on the $\overline{\Phi}_a$-plane, where point A is (x, jy). This point can be expressed in terms of the angle θ,

$$x = \text{OA} \cos \theta$$

$$y = \text{OA} \sin \theta$$

Substituting these values into equation 8.12, we obtain

$$\text{OA}^2(\cos^2 \theta+\sin^2 \theta)+\text{OA} \, . \, 2 \cos \theta \frac{M^2}{M^2-1}+\frac{M^2}{M^2-1} = 0 \tag{8.22}$$

Solving equation 8.22 for OA, we have

$$\text{OA} = -\cos \theta \frac{M^2}{M^2-1} \pm \sqrt{\left[\left(\frac{\cos \theta M^2}{M^2-1}\right)^2 - \frac{M^2}{M^2-1}\right]} \tag{8.23}$$

To plot the M contour on the Nichols diagram, a value of M is specified and OA is plotted for a series of values of the angle θ, using equation 8.23. The value of M is indicated on the contour in decibels. (For example, if M = 0·5, 1·0, 2·0 and 4·0, the respective values of M in decibels are −6, 0, 6 and 12.)

The Nichols diagram is shown calibrated for M.

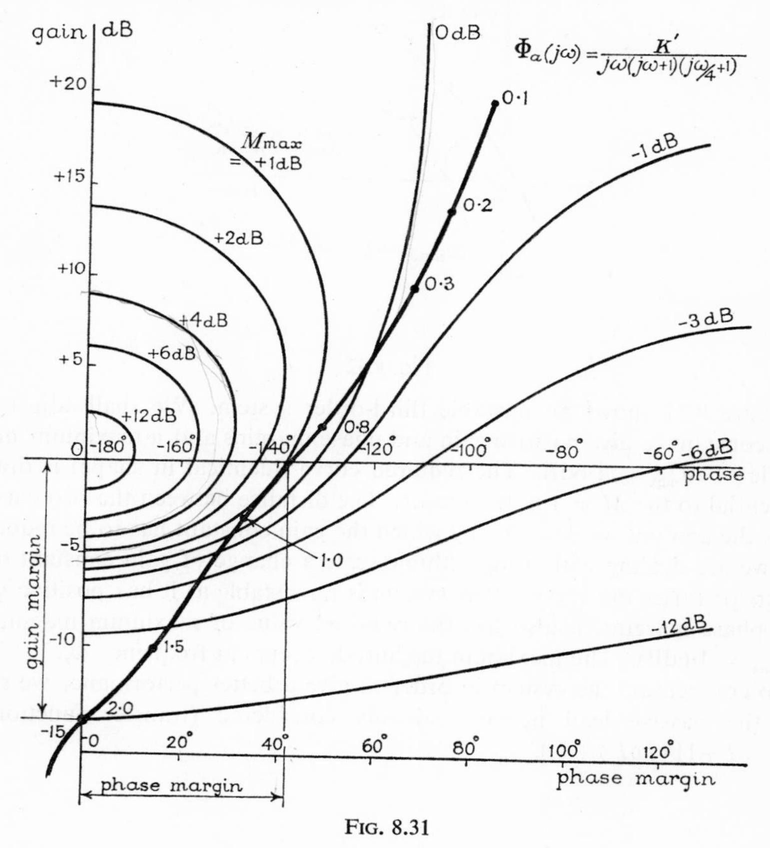

FIG. 8.31

Figure 8.31. A phase margin scale is shown on the figure. The gain and phase margins for the example are also shown on the figure. As they are positive the system is stable for a unity gain constant. Only the portion of the graph required for the system analysis is shown. The graph has also been calibrated for frequency ω.

In the previous section we found that the M circle tangential to the Nyquist locus gives the maximum magnitude M_{max}. The frequency at the point where the M circle is tangential is the resonant frequency ω_r. The same applies to the Nichols diagram. A system's transient and steady-state performance can be improved by adjusting the value of M_{max}. The resonant frequency ω_r gives an indication of a system's response.

Let us consider the gain adjustment and compensation of an unstable third-order system.

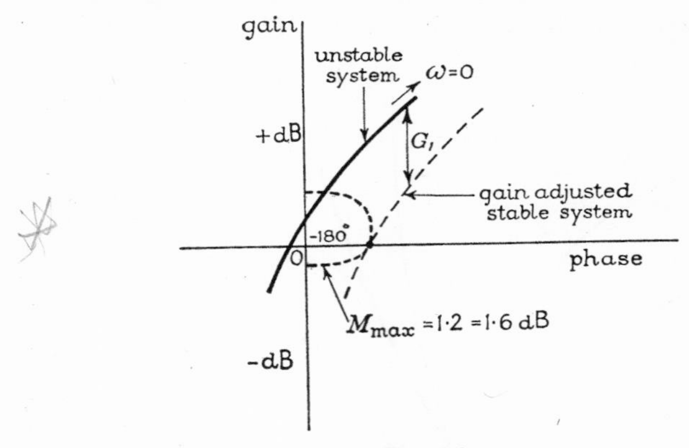

FIG. 8.32

Figure 8.32 shows an unstable third-order system. We shall adjust the gain constant to give positive gain and phase margins and a maximum magnitude of M_{max} = 1·6dB. The required curve (identical in shape) is drawn tangential to the M = 1·6dB contour. The distance between the two curves, G_1 is the amount, in decibels, by which the gain constant has to be reduced. (As we are dealing with a logarithmic scale a change of gain constant only lowers or raises the curve.) The system is now stable as it has positive gain and phase margins, it also has the required value of maximum magnitude (M_{max} = 1·6dB). The maximum magnitude occurs at frequency ω_r.

To compensate the system in order to give a better performance we shall use the passive lead network already considered (transfer function = $(A)(j\omega T+1)/(j\omega TA+1)$).

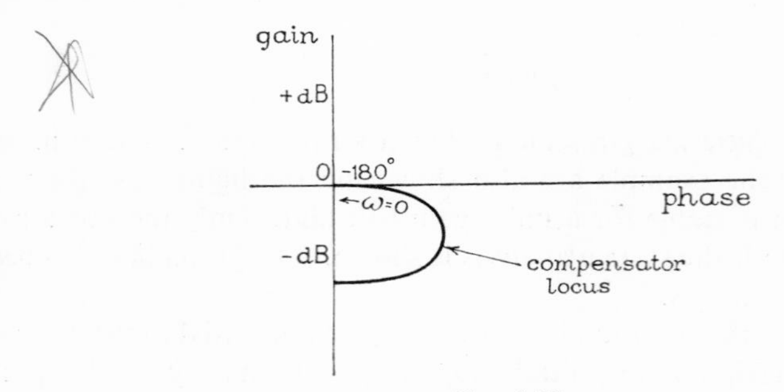

FIG. 8.33

Figure 8.33 shows the Nichols diagram of the compensating network. To find the effect of the compensation on the adjusted curve in figure 8.31, the curve of the compensating network and the system curve are drawn on the same axes. The resulting curve is found by adding the gains and the phase angles at the same frequency.

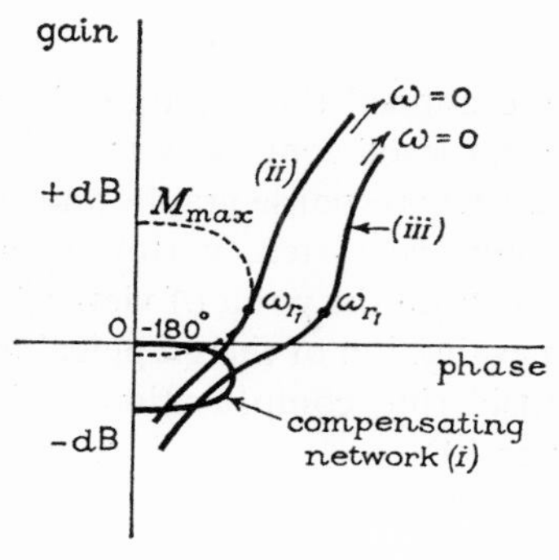

FIG. 8.34

Figure 8.34. This figure shows (*i*) the gain-phase curve of the passive lead compensation network, (*ii*) the gain-phase curve of the system adjusted for gain from figure 8.32 and (*iii*) the resulting gain-phase curve found by adding the gains and the phases of curves (*i*) and (*ii*) at the same frequency. To illustrate the plotting of curve (*iii*) we shall take a particular frequency ω_1. For this frequency the gain obtained for curve (*iii*), from curves (*i*) and (*ii*), is $G_1 - G_2$ and the phase is $\theta_1 - \theta_2$.

To obtain the required $M_{\max}$ the gain constant has to be readjusted. The resulting resonant frequency ω_r will be much higher than without the passive lead compensation, and this results in a better response.

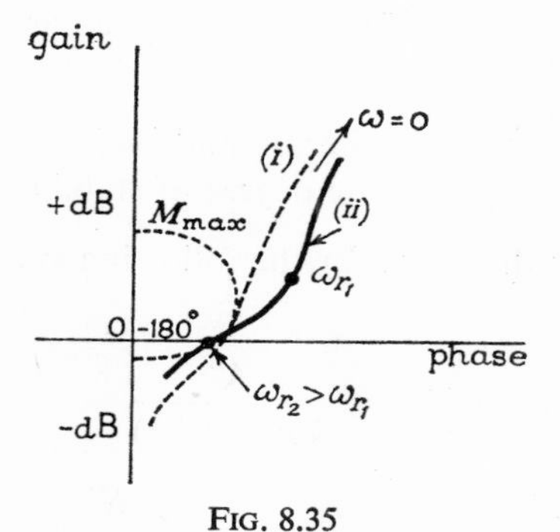

FIG. 8.35

Figure 8.35 shows the uncompensated gain-phase curve (*i*) and the compensated gain-phase curve (*ii*). Both curves are adjusted for the required maximum magnitude $M_{\max}$. This figure indicates that passive lead compensation gives a higher resonant frequency and yet at lower frequencies has little effect.

8.8 Conclusion

In this chapter we have examined three graphical methods of showing the open-loop frequency response of control systems. We have discussed the interpretation of the frequency response graphs and the change of the graphs with compensation to improve system performance. A summary of the frequency response graphs from the point of view of system stability is given in the next chapter. An evaluation of the graphical methods and their interpretation, especially considering compensation to improve system performance, is given in chapter 10.

8.9 Examples

1. Find the transfer functions, as functions of frequency, of the systems shown in the following figures:

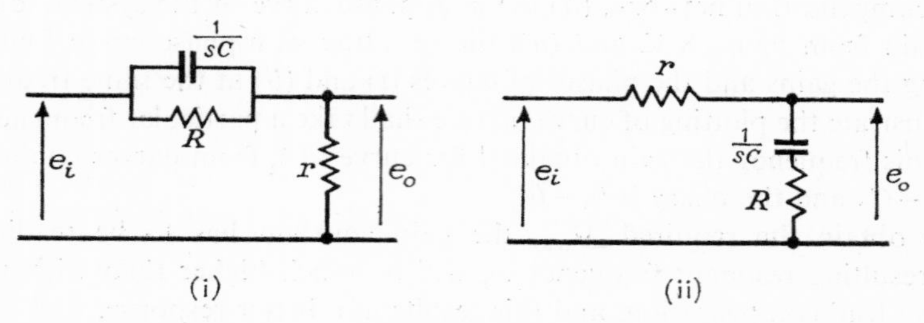

where e_i and e_o are the input and output voltages respectively.

Explain why these transfer functions are a complete description of the systems.

2. State the Nyquist stability criterion and indicate how the Nyquist locus of a system shows the approximate degree of stability of a system.

3. Draw the Nyquist diagrams for the following system open-loop transfer functions:

$$(i)\quad \Phi_a(j\omega) = \frac{K'}{j\omega(j\omega 2+1)}$$

$$(ii)\quad \Phi_a(j\omega) = \frac{K'}{j\omega(j\omega 0{\cdot}5+1)(j\omega 0{\cdot}33+1)}$$

$$(iii)\quad \Phi_a(j\omega) = \frac{K'(j\omega+1)}{j\omega(j\omega 0{\cdot}5+1)}$$

$$(iv)\quad \Phi_a(j\omega) = \frac{K'}{(j\omega)^2(j\omega 4+1)}$$

$$(v) \quad \Phi_a(j\omega) = \frac{K'(j\omega 100+1)}{j\omega(j\omega 4+1)(j\omega 50+1)}$$

4. Find the range of gain constant K for which the systems in question 3 are stable.

(all K; $0 < K < 30$; all K; no K; no K)

5. The diagrams show part of the Nyquist loci for four control systems; comment on the stability of each system.

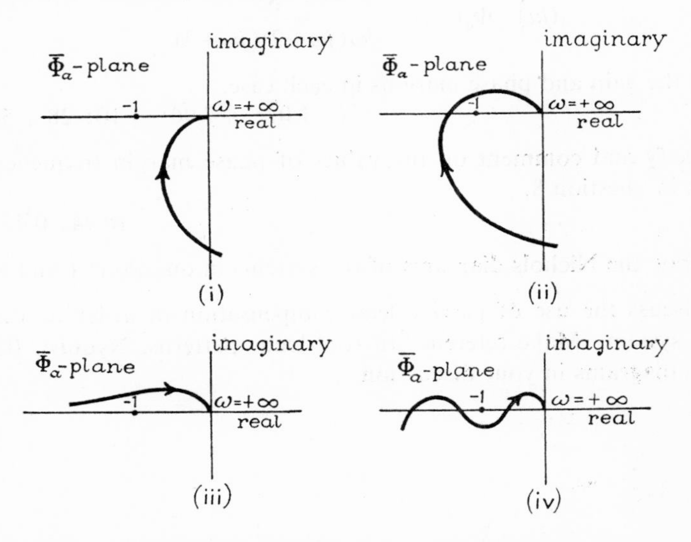

Give examples of possible open-loop transfer functions $\Phi_a(j\omega)$ in each case.

6. A control system has an open-loop transfer function

$$\Phi_a(j\omega) = \frac{1}{j\omega(j\omega+1)(j\omega+2)}$$

Draw the Nyquist diagram for this system and find the gain constant K for a maximum magnitude $M_{max} = 2$. Determine the system resonant frequency ω_r at this maximum magnitude.

(2; 0·8)

7. A remote position control system has a damping ratio $\zeta = 0{\cdot}6$ and an undamped natural frequency of $\omega_n = 15$ radians per second. If the system characteristic equation is second-order, draw its Nyquist diagram and find the maximum magnitude and resonant frequency of the system.

(1·042; 7·94)

8. Draw the Bode diagrams for the following control system open-loop transfer functions;

$$(i)\quad \Phi_a(j\omega) = \frac{1}{j\omega(j\omega+0\cdot5)}$$

$$(ii)\quad \Phi_a(j\omega) = \frac{1}{j\omega(j\omega+1)(j\omega0\cdot5+1)}$$

$$(iii)\quad \Phi_a(j\omega) = \frac{5}{j\omega(j\omega+2)(j\omega+3)}$$

Specify the gain and phase margins in each case.

(+1·07; +7·96; +10; 28°; 52°; 57°)

9. Specify and comment on the values of phase margin frequency of the systems in question 8.

(0·94; 0·45; 0·833)

10. Draw the Nichols diagrams of the systems in questions 1 and 8.

11. Discuss the use of passive lead compensation in order to stabilize a control system. Make reference to root-locus patterns, Nyquist, Bode and Nichols diagrams in your discussion.

CHAPTER 9

System Stability

*9.1 Introduction

In previous chapters we have discussed the idea of stability with reference to various descriptions of control systems. We now shall consider the conditions and criteria of stability. Firstly we shall examine three areas of stability: the stable system, the unstable system and the conditionally stable system. Consider a unit step function input to a system. The system is stable if the initial transients of the output die out and the output becomes a constant value after some time. If, however, the output increases indefinitely, or oscillates at a constant amplitude, the system is unstable. A system is considered conditionally stable if, with the alteration of the gain constant, the system is stable only over a range of gain constant.

9.2 Stability from root-locus patterns and frequency response graphs

Root-locus patterns

In chapter 7 we found that a system is unstable if any roots of the characteristic equation have positive real parts and if conjugate complex roots exist on the imaginary axis (this indicates oscillations of a constant amplitude). The root-locus pattern is the path of the closed-loops pole for a variation of the gain constant of a system.

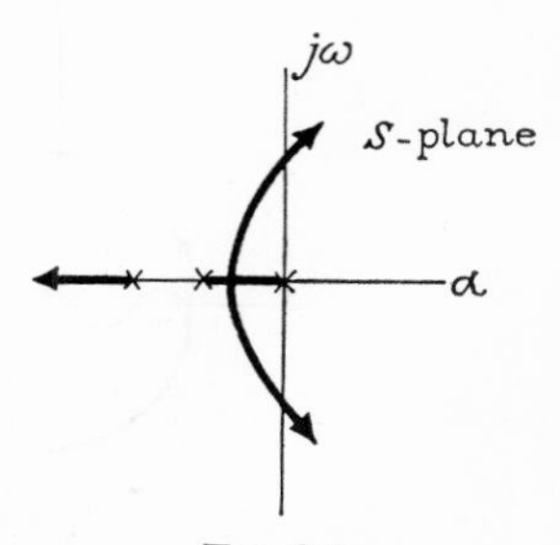

FIG. 9.1

Figure 9.1. This figure shows the root-locus pattern of a third-order system. If a value of gain constant K takes the root-locus past the imaginary axis into the right side of the s-plane that value of gain constant makes the system unstable.

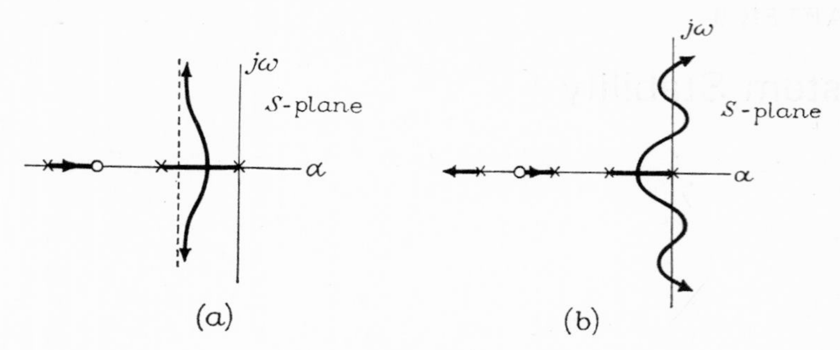

FIG. 9.2

Figure 9.2(a) shows the root-locus pattern of a system which is stable for all values of gain constant. Figure 9.2(b) shows part of the root-locus pattern of a system which is conditionally stable. The root-locus pattern of a conditionally stable system crosses the imaginary axis three times showing definite 'ranges' of gain constant over which the system is stable.

Nyquist diagrams

In chapter 8 we found that the enclosure of the $(-1, j0)$ point by the Nyquist locus (locus of the open-loop frequency response on the $\bar{\Phi}_a$-plane) indicates that a system is unstable. This is the Nyquist criterion of stability.

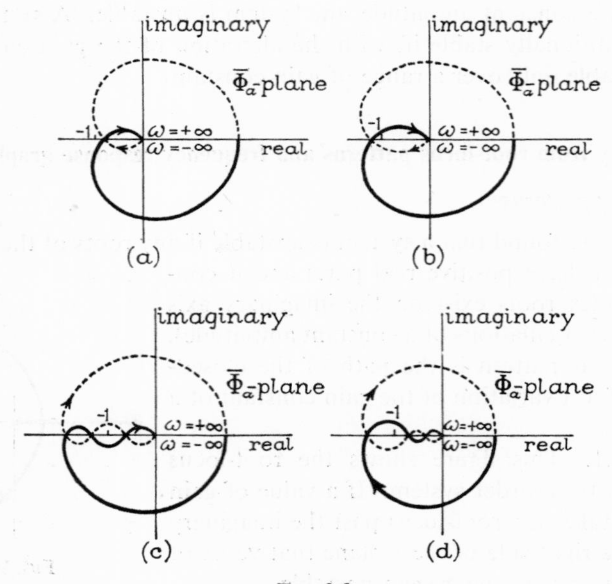

FIG. 9.3

Figure 9.3 (*a*) shows the Nyquist diagram of a stable system. Figure 9.3 (*b*) shows the Nyquist diagram of an unstable system. Figure 9.3 (*c*) shows the Nyquist diagram of a conditionally stable system with the gain constant such that the system is stable. (The locus does not enclose the $(-1, j0)$ point.) Figure 9.3 (*d*) shows the Nyquist diagram of a conditionally stable system with the gain constant such that the system is unstable. Figures 9.3 (*c*) and (*d*) clearly indicate that the system has definite 'ranges' of gain constant over which it is stable.

The root-locus pattern shows the values of gain constant over which a system is stable, whereas the Nyquist diagram just indicates whether the system is stable or unstable for a specific gain constant. (Figures 9.3 (*a*) and (*b*) are the same system with different gain constants.) However, the gain margin and phase margin on the Nyquist diagram give a guide to the approach to stability, or instability, of a system for a particular gain constant. Negative gain and phase margins indicate an unstable system, positive gain and phase margins indicate a stable system.

Bode diagrams

The Bode diagrams consist of two frequency response graphs which provide similar information to Nyquist diagrams, but are often easier to interpret and manipulate. Phase and gain margins are shown on the Bode diagrams and they are indicative of stability in the same way as for Nyquist diagrams.

Nichols diagrams

The Nichols diagram is one locus combining the two graphs of the Bode diagrams. Phase and gain margins are shown on the Nichols diagrams and they are indicative of stability in the same way as for Nyquist diagrams.

9.3 The Hurwitz-Routh condition

A system characteristic equation (denominator of the closed-loop transfer function) describes a system which is either stable or unstable. In general the characteristic equation is

$$b_0 + b_1 s + b_2 s^2 + b_3 s^3 + \dots + b_n s^n = 0 \qquad (9.1)$$

where $b_0, b_1, b_2 \dots b_n$ are constant real coefficients and s is the complex variable $\alpha + j\omega$. We know that if the roots of equation 9.1 have a positive real part the system described by such a characteristic equation is unstable. It is possible to find out whether a characteristic equation has roots with positive real parts, or not, by examination of the coefficients of the equation. A. Hurwitz and E. J. Routh independently developed the same method of finding the condition under which equations have roots with only negative real parts. The condition, using the coefficients of equation 9.1, can be expressed in the form of determinants, as follows

$$\begin{vmatrix} b_{n-1} & b_n & 0 & 0 & \cdot & \cdot & \cdot & \cdot & \cdot \\ b_{n-3} & b_{n-2} & b_{n-1} & b_n & 0 & \cdot & \cdot & \cdot & \cdot \\ b_{n-5} & b_{n-4} & \cdot & \cdot & \cdot & \cdot & \cdot & \cdot & \cdot \\ \cdot & \cdot & \cdot & \cdot & \cdot & \cdot & \cdot & \cdot & \cdot \\ \cdot & \cdot & \cdot & \cdot & \cdot & \cdot & \cdot & \cdot & \cdot \\ \cdot & \cdot & \cdot & \cdot & \cdot & \cdot & \cdot & b_5 & b_6 \\ \cdot & \cdot & \cdot & \cdot & b_0 & b_1 & b_2 & b_3 & b_4 \\ \cdot & \cdot & \cdot & \cdot & \cdot & 0 & b_0 & b_1 & b_2 \\ \cdot & \cdot & \cdot & \cdot & \cdot & \cdot & 0 & 0 & b_0 \end{vmatrix} > 0 \qquad (9.2)$$

Determinant inequality 9.2 is the first condition. The second condition is the determinant inequality obtained by removing the last column and the last row of the determinant inequality 9.2,

$$\begin{vmatrix} b_{n-1} & b_n & 0 & 0 & \cdot & \cdot & \cdot & \cdot \\ b_{n-3} & b_{n-2} & b_{n-1} & b_n & 0 & \cdot & \cdot & \cdot \\ b_{n-5} & b_{n-4} & b_{n-3} & \cdot & \cdot & \cdot & \cdot & \cdot \\ \cdot & \cdot & \cdot & \cdot & \cdot & \cdot & \cdot & \cdot \\ \cdot & \cdot & \cdot & \cdot & \cdot & \cdot & \cdot & \cdot \\ \cdot & \cdot & \cdot & \cdot & \cdot & \cdot & \cdot & \cdot \\ \cdot & \cdot & \cdot & 0 & b_0 & b_1 & b_2 & b_3 \\ \cdot & \cdot & \cdot & \cdot & 0 & 0 & b_0 & b_1 \end{vmatrix} > 0 \qquad (9.3)$$

Determinant inequality 9.3 is the second condition. The further conditions are found by subsequent removal of the last row and last column of each determinant inequality. The last two conditions are

$$\begin{vmatrix} b_{n-1} & b_n \\ b_{n-3} & b_{n-2} \end{vmatrix} > 0 \quad \text{and} \quad b_{n-1} > 0$$

To illustrate this stability condition we shall consider the example of a system which has the following characteristic equation:

$$2+2s+4s^2+3s^3+s^4 = 0 \qquad (9.4)$$

$$b_{n-4} = 2,\ b_{n-3} = 2,\ b_{n-2} = 4,\ b_{n-1} = 3,\ b_n = 1$$

We shall use the Hurwitz-Routh conditions to examine whether equation 9.4 has roots with negative real parts only. If the conditions hold the system is stable.

Condition 1 is the determinant inequality,

$$\begin{vmatrix} 3 & 1 & 0 & 0 \\ 2 & 4 & 3 & 1 \\ 0 & 2 & 2 & 4 \\ 0 & 0 & 0 & 2 \end{vmatrix} > 0$$

which becomes

$$2\begin{vmatrix} 3 & 1 & 0 \\ 2 & 4 & 3 \\ 0 & 2 & 2 \end{vmatrix} > 0$$

Evaluating this determinant inequality we have $4 > 0$, therefore, condition 1 holds.

Condition 2 is the determinant inequality obtained by removing the last column and the bottom row of the determinant in condition 1,

$$\begin{vmatrix} 3 & 1 & 0 \\ 2 & 4 & 3 \\ 0 & 2 & 2 \end{vmatrix} > 0$$

Evaluating this determinant inequality we have $2 > 0$, therefore, condition 2 holds.

Condition 3 is the determinant inequality obtained by removing the last column and the bottom row of the determinant in condition 2,

$$\begin{vmatrix} 3 & 1 \\ 2 & 4 \end{vmatrix} > 0$$

Evaluating this determinant inequality we have $10 > 0$, therefore, condition 3 holds.

Condition 4 is the inequality obtained by removing the last column and bottom row of the determinant in condition 3, $3 > 0$, therefore, condition 4 holds.

All the conditions hold, therefore equation 9.4 has only roots with negative real parts; hence, it represents the characteristic equation of a stable system.

We can see that the conditions would not have held if any of the coefficients of equation 9.4 were negative, or missing. This is true in general; if a system has negative or missing coefficients in its characteristic equation then an unstable system is indicated. However, all positive coefficients present in a characteristic equation does not necessarily indicate a stable system, and the Hurwitz-Routh condition can be applied to quickly determine whether the system is stable.

9.4 Examples

1. Discuss how the root-locus pattern, Nyquist, Bode and Nichols diagrams indicate the degree of stability of a control system.

2. A remote position control system has the closed-loop transfer function

$$\overline{\Phi} = \frac{K}{s^2+2\sqrt{K}s+K}$$

Show by means of the Hurwitz-Routh condition that the system is stable for all values of gain constant K.

3. Use the Hurwitz-Routh condition to find whether the following systems are stable,

(*i*) open-loop transfer function $\overline{\Phi}_a = \dfrac{40}{s(s+2)(s+3)}$

(*ii*) closed-loop transfer function $\overline{\Phi} = \dfrac{10}{s^2+3s+6}$

(*iii*) open-loop transfer function $\overline{\Phi}_a = \dfrac{10}{s^2(s+2)}$

(No; Yes; No)

4. A servo-mechanism has an open-loop transfer function

$$\overline{\Phi}_a = \frac{K(s+1)}{s^2(s+2)}$$

Use the Hurwitz-Routh condition to find the range of the gain constant K for which the servo-mechanism is stable.

($K > \frac{1}{2}$)

5. A control system has the following characteristic equation,

$$s^3+3s^2-as+1 = 0,$$

where a is a constant coefficient. Find the range of the coefficient a for which the system is stable.

(a is negative and $a > -\frac{1}{3}$)

6. Explain, with reference to root-locus patterns, Nyquist and Nichols diagrams, what is meant by a conditionally stable system.

CHAPTER 10

Compensation

10.1 Introduction

In chapters 7 and 8 we have discussed the analysis of control systems and briefly touched on the modification of system performance. We shall now consider the improvement of performance of simple closed-loop systems, with reference to the transient analysis (root-locus patterns) and the frequency response analysis (Nyquist, Bode and Nichols diagrams) of chapters 7 and 8 respectively.

The first considerations in designing a control system are that the system should be stable and give an output that is accurate within certain specified limits. In practice they may be a compromise between desired and possible limits. Thus we require design specifications. The first steps in designing the system are to suggest possible limits, i.e. the output power required, the response time and some suggestion of accuracy with reference to an input. This will give a guide to the selection of system components, which can be built up into a rough experimental system (or simulated on an analogue computer). The system may be found to be unstable, but a gain adjustment may be all that is required for stability. However, it is possible that stability can only be obtained by the addition of further components, such as a tachometer for velocity feed-back, or other devices depending on the type of system. Although a system has been made stable, it does not necessarily follow that it will meet the output requirements. Therefore further alterations, usually in the form of additional components, are required.

Analysis of the system being designed reveals its shortcomings, and this is often called *design by analysis*. We shall consider electrical networks for the improvement of a system performance. These are called *compensating networks* and their insertion into a system is called *compensation*.

In order to improve a system's performance we must be able to specify its performance in terms of various defining quantities. Specifications in the time domain can be in terms of the system response to a unit step function input.

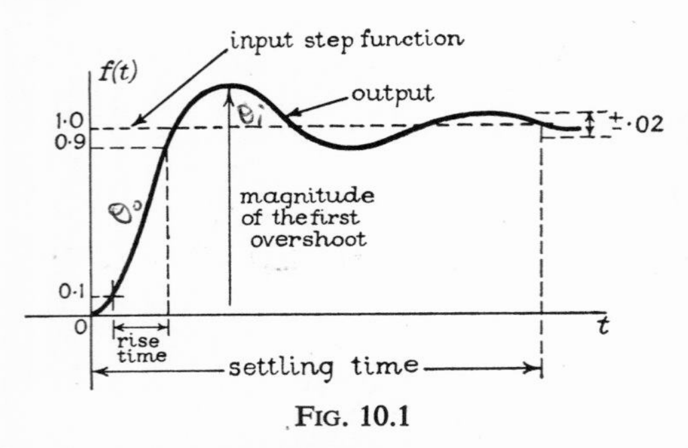

FIG. 10.1

Figure 10.1 shows the response of a system to a unit step function. It can be appreciated in terms of:

(*i*) The time taken for the output to rise from 10% to 90% of the value of the input. This is called the *rise time*.

(*ii*) The magnitude of the first overshoot (unless there is no overshoot).

(*iii*) The time taken for the output to reach and remain within 2% of its final (steady-state) value.

When considering systems with dominant conjugate complex roots, which can be compared with a second-order system, the damping ratio ζ and the undamped natural frequency ω_n give an indication of performance. ζ gives an indication of the degree of system stability and ω_n the speed of response.

In terms of frequency response, the gain and phase margins indicate the degree of system stability (they can be compared with ζ). Maximum magnitude M_{max} indicates the amount of overshoot, and hence gives an estimation of the damping ratio ζ. The frequency, on the Nyquist diagram, at which M_{max} occurs is the system resonant frequency ω_r and is a good guide to the speed of response.

It is also desirable to specify a system in terms of its steady-state error. To do this we must categorize control systems into 'types'.

10.2 Types of control systems

In general a unity feed-back single-loop system can be represented by a simple block diagram.

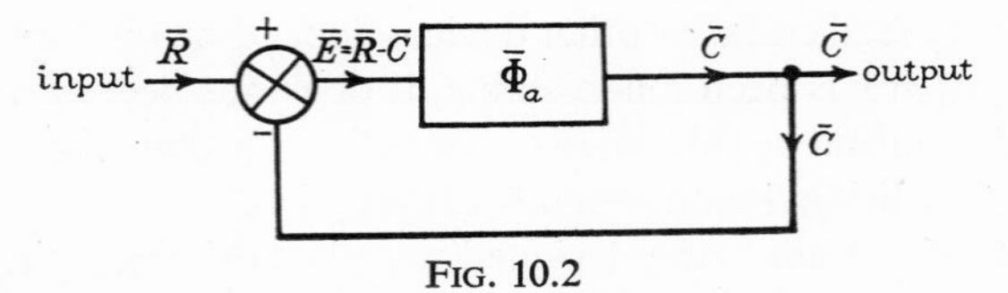

FIG. 10.2

Figure 10.2 shows the system. It has an open-loop transfer function,

$$\bar{\Phi}_a = \frac{\bar{C}}{\bar{E}} \qquad (10.1)$$

We shall classify a system, in terms of a reference input, according to its steady-state error. Expressing equation 10.1 in terms of the error $\bar{E}(=\bar{R}-\bar{C})$, where R is the reference input, we have

$$\frac{\bar{R}-\bar{E}}{\bar{E}} = \bar{\Phi}_a$$

therefore

$$\bar{E} = \frac{\bar{R}}{1+\bar{\Phi}_a} \tag{10.2}$$

We have seen that the open-loop transfer function can, in general, be expressed as

$$\bar{\Phi}_a = \frac{K(s+a_1)(s+a_2)(s+a_3)\ldots(s+a_m)}{s^n(s+b_1)(s+b_2)(s+b_3)\ldots(s+b_p)} \tag{10.3}$$

where K is the gain constant, $a_1, a_2, a_3 \ldots a_m$ and $b_1, b_2, b_3 \ldots b_p$ are constant coefficients, and $n = 0, 1, 2, 3$, etc. (the order of the denominator is greater than the order of the numerator). Equation 10.3 will, in general, be written as

$$\bar{\Phi}_a = \frac{K\bar{N}}{s^n\bar{d}} \tag{10.4}$$

Substituting equation 10.4 into 10.2 we obtain

$$\bar{E} = \frac{\bar{R}}{1+K\bar{N}/s^n\bar{d}} \tag{10.5}$$

We shall use equation 10.5 to obtain a classification of systems according to type. Consider three reference inputs: the unit step function, the unit ramp function and the unit parabolic function.

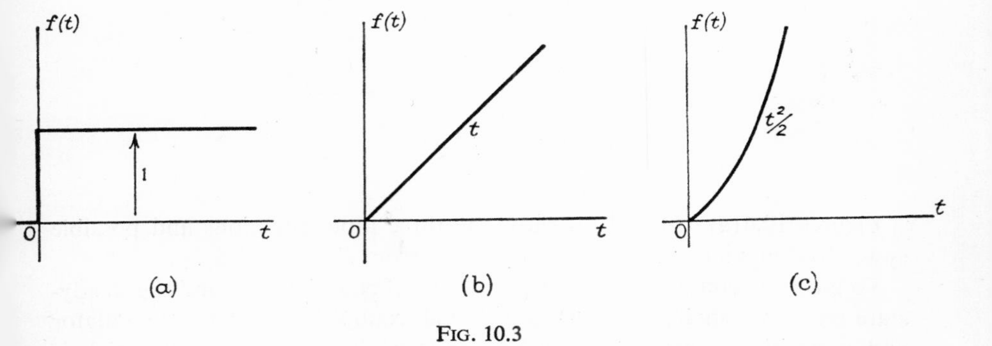

FIG. 10.3

Figures 10.3(*a*), (*b*) and (*c*) show respectively the unit step function $\mathscr{L}f(t) = 1/s$, the unit ramp function $\mathscr{L}f(t) = 1/s^2$, and the unit parabolic function $\mathscr{L}f(t) = 1/s^3$.

The unit step function input to a system is an input 'position' and in the steady state the system output should be a 'position'. If the system has a steady-state error for a unit step function input, it is a *position error* E_p.

The unit ramp function input to a system is an input 'velocity' and in the steady state the system output should be a 'velocity'. If the system has a steady-state error for a unit ramp function, it is a *velocity error* E_v.

The unit parabolic function input to a system is an input 'acceleration' and in the steady state the system output should be an 'acceleration'. If the system has a steady-state error for a unit parabolic function, it is called an *acceleration error* E_a.

The terms 'position', 'velocity' and 'acceleration' are used for convenience, even when considering systems which are not dynamic.

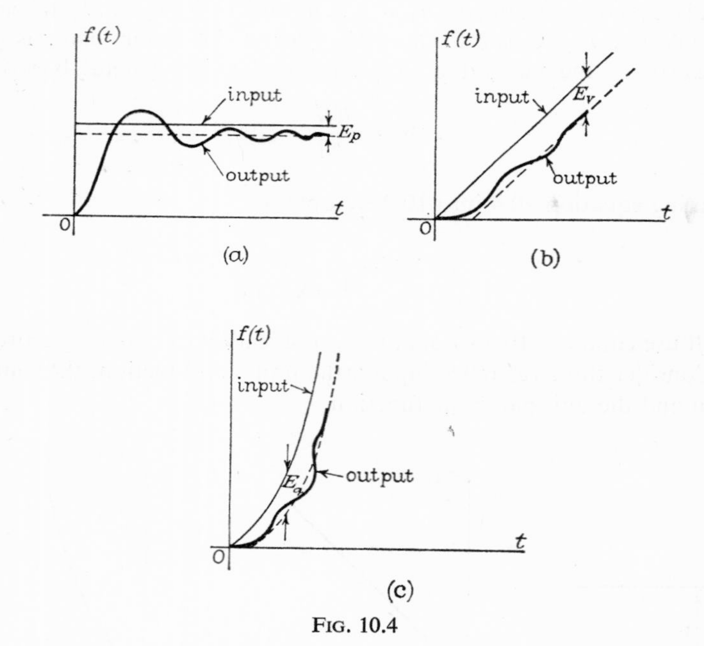

FIG. 10.4

Figures 10.4(*a*), (*b*) and (*c*) show the three input functions and possible system outputs indicating the steady-state errors E_p, E_v and E_a.

To illustrate control system types in terms of equation 10.5 and the steady-state errors we shall consider the practical examples of a voltage regulator and a position controller. A regulator is a closed-loop system which is designed to give a constant output for a constant input.

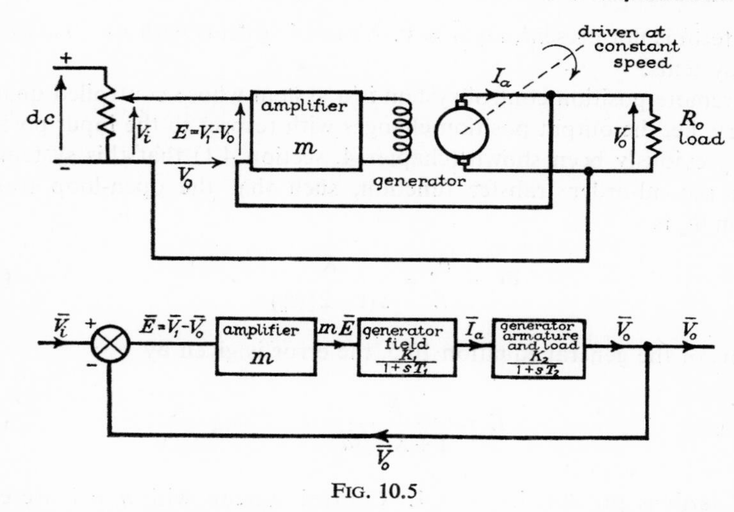

FIG. 10.5

Figure 10.5 shows a voltage regulator and its block diagram. T_1 is the field time constant ($T_1 = L_f/R_f$, where L_f and R_f are the field inductance and resistance, respectively). T_2 is the armature and load time constant ($T_2 = L_a/(R_a+R_o)$ where L_a, R_a and R_o are the armature inductance, resistance and load resistance, respectively). m, K_1 and K_2 are the amplifier gain, the generator field constant and armature constant.

The voltage regulator transfer function for a resistive load is given by

$$\frac{\overline{V}_o}{\overline{V}_i} = \frac{\overline{\Phi}_a}{1+\overline{\Phi}_a}$$

where $\overline{\Phi}_a$ is the open-loop transfer function, therefore

$$\frac{\overline{V}_o}{\overline{V}_i} = \frac{\dfrac{mK_1K_2}{(1+sT_1)(1+sT_2)}}{1+\dfrac{mK_1K_2}{(1+sT_1)(1+sT_2)}}$$

The open-loop transfer function is

$$\overline{\Phi}_a = \frac{\overline{V}_o}{\overline{E}} = \frac{mK_1K_2/T_1T_2}{(s+1/T_1)(s+1/T_2)} \tag{10.6}$$

where $(mK_1K_2)/(T_1T_2)$ is the system gain constant. In terms of the general equation 10.5, the error is given by

$$\overline{E} = \frac{\overline{R}}{1+K\overline{N}/\overline{d}} \tag{10.7}$$

The s^n term is not present, i.e. $n = 0$. Control systems with $n = 0$ are called *type* 0 systems.

The remote position control system is a system whose controlled quantity can vary, i.e. the output position changes with respect to the input position. It has previously been shown (chapter 4, section 4.1) that this system can have a second-order transfer function, such that the open-loop transfer function $\bar{\Phi}_a$ is

$$\bar{\Phi}_a = \frac{\bar{\theta}_o}{\bar{E}} = \frac{K}{s(s+2\zeta\omega_n)} \tag{10.8}$$

In terms of the general equation 10.5, the error is given by

$$\bar{E} = \frac{\bar{R}}{1+K\bar{N}/s\bar{d}} \tag{10.9}$$

The s^n term is present, i.e. $n = 1$. Control systems with $n = 1$ are called *type* 1 systems.

Thus control systems are classified according to the value of n in the general open-loop transfer function $\bar{\Phi}_a = (K\bar{N})/(s^n\bar{d})$. (The remote position control system would be a type 2 system if acceleration feed-back had been employed.)

Evaluation of the steady-state errors

To find the steady-state errors E_p, E_v and E_a it is necessary to find the steady-state solution of equation 10.5. The Laplace final-value theorem enables us to do this.

The Laplace final-value theorem

This theorem enables a control system steady-state error time solution to be found (i.e. the solution at time $t = \infty$) from the system error equation without performing the inverse transformation. It states that for a function of time $f(t)$, where $\mathscr{L}f(t) = f(s)$, the value of $f(t)$ as time t tends to infinity is given by $sf(s)$, where the complex variable s tends to zero. That is,

$$\underset{t\to\infty}{f(t)} = \underset{s\to 0}{sf(s)} \tag{10.10}$$

This, however, is only true for the solutions of stable systems; unstable systems will not have a final (constant) value of error.

The error constants

We shall obtain expressions for the error constants E_p, E_v and E_a for the systems type 0, 1 and 2. The error in its most general form is

$$\bar{E} = \frac{\bar{R}}{1+\dfrac{K(s+a_1)(s+a_2)\ldots(s+a_m)}{s^n(s+b_1)(s+b_2)\ldots(s+b_p)}}$$

(*i*) Position Error Constant E_p

This can be evaluated with a unit step function input, $\bar{R} = 1/s$, and using the Laplace final-value theorem

$$E_p = \left[\frac{s\,.\,1/s}{1+\dfrac{K(s+a_1)(s+a_2)\ldots(s+a_m)}{s^n(s+b_1)(s+b_2)\ldots(s+b_p)}}\right]_{s\to 0} \tag{10.11}$$

For a type 0 system ($n = 0$) equation 10.11 gives

$$E_p = \frac{1}{1+K_p} \tag{10.12}$$

where $K_p = \dfrac{K(a_1\,a_2\,a_3\ldots a_m)}{(b_1\,b_2\,b_3\ldots b_p)}$

For systems of type 1 and above ($n = 1, 2, 3$, etc.) equation 10.11 is zero, i.e. there is no position error constant.

(*ii*) Velocity Error Constant E_v

This can be evaluated with a unit ramp function input, $\bar{R} = 1/s^2$, and using the Laplace final-value theorem.

$$E_v = \left[\frac{s\,.\,1/s^2}{1+\dfrac{K(s+a_1)(s+a_2)\ldots(s+a_m)}{s^n(s+b_1)(s+b_2)\ldots(s+b_p)}}\right]_{s\to 0} \tag{10.13}$$

For a type 0 system ($n = 0$) equation 10.13 is infinite.
For a type 1 system ($n = 1$) equation 10.13 gives

$$E_v = \frac{1}{K_v} \tag{10.14}$$

where $K_v = \dfrac{K(a_1\,a_2\,a_3\ldots a_m)}{(b_1\,b_2\,b_3\ldots b_p)}$

For systems of type 2 and above ($n = 2, 3, 4$, etc.) equation 10.13 is zero, i.e. there is no velocity error constant.

(*iii*) Acceleration Error Constant E_a

This can be evaluated with a unit parabolic function input $\bar{R} = 1/s^3$ and using the Laplace final-value theorem.

$$E_a = \left[\frac{s \,.\, 1/s^3}{1+\dfrac{K(s+a_1)(s+a_2)\ldots(s+a_m)}{s^n(s+b_1)(s+b_2)\ldots(s+b_p)}}\right]_{s\to 0} \qquad (10.15)$$

For systems type 0 and 1 ($n = 0$ and 1) equation 10.15 is infinite. For a type 2 system ($n = 2$) equation 10.15 gives

$$E_a = \frac{1}{K_a} \qquad (10.16)$$

where

$$K_a = \frac{K(a_1\, a_2\, a_3 \ldots a_m)}{(b_1\, b_2\, b_3 \ldots b_p)} \qquad (10.17)$$

For systems type 3 and above ($n = 3, 4, 5$, etc.) equation 10.15 is zero, i.e. there is no acceleration error constant.

The constants K_p, K_v and K_a refer to the open-loop transfer function and re often called the steady-state error coefficients.

Summary

System type	Steady-state error
0	Constant position error $E_p = \dfrac{1}{1+K_p}$ Infinite velocity and acceleration errors
1	Constant velocity error $E_v = \dfrac{1}{K_v}$ Zero position error Infinite acceleration error
2	Constant acceleration error $E_a = \dfrac{1}{K_a}$ Zero position and velocity errors
	Error Coefficient K'
0	position $[K_p]_{t\to\infty} = [\Phi_a]_{s\to 0}$
1	velocity $[K_v]_{t\to\infty} = [s\Phi_a]_{s\to 0}$
2	acceleration $[K_a]_{t\to\infty} = [s^2\Phi_a]_{s\to 0}$

A system's performance can be partly assessed in terms of its steady-state error, the most common type being the velocity error.

10.3 Feed-forward compensation

We shall consider the compensation of unity feed-back control systems. The compensator modifies the system error.

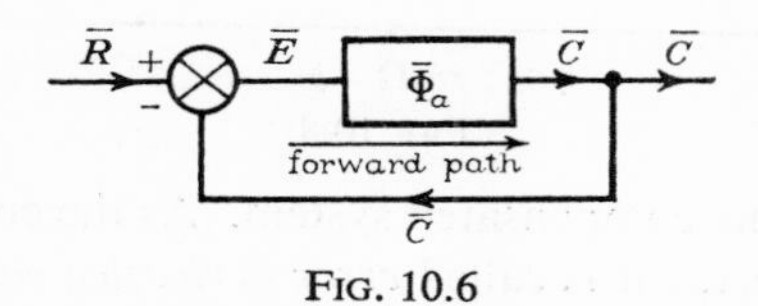

FIG. 10.6

Figure 10.6 shows the block diagram of the system to be improved. The compensator is to be inserted in the forward path.

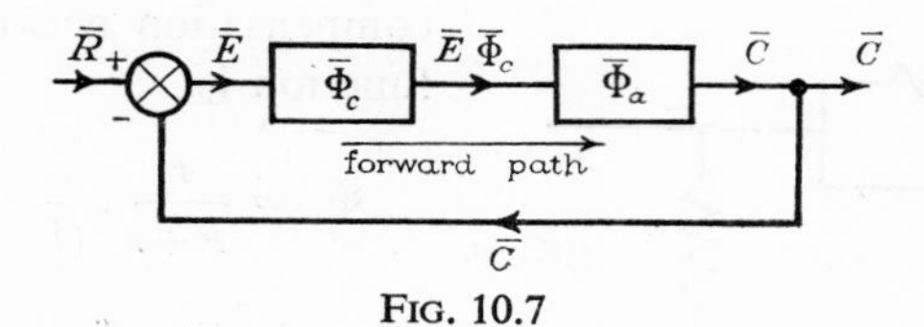

FIG. 10.7

Figure 10.7 shows the block diagram of the improved system. Feed-forward compensation is obtained by modifying the error E with the compensation device which has a transfer function $\bar{\Phi}_c$.

Error derivative compensation

Let us consider a system which has a small steady-state error, but whose output responds more slowly to an input than is desired, i.e. the system transient response is not good enough. We shall use feed-forward compensation to improve the system's performance. The effect of the error E, through the system, must be made faster. We therefore modify the error by adding to it an estimation of its rate of change. It will then anticipate changes in the system and response will be faster. The new error signal will be the original signal E plus a signal proportional to the rate of change of the original error, i.e. proportional to $dE/dt = sE$. The modified error can be expressed as

$$\bar{E} + T_1 s\bar{E} = \text{modified error}$$

(original error) (signal proportional to the rate of change of the error)

therefore

$$\text{modified error} = \bar{E}(1+T_1 s) \tag{10.18}$$

where T_1 is a constant. Equation 10.18 indicates that the compensator transfer function $\bar{\Phi}_c$ should be $(1+T_1 s)$.

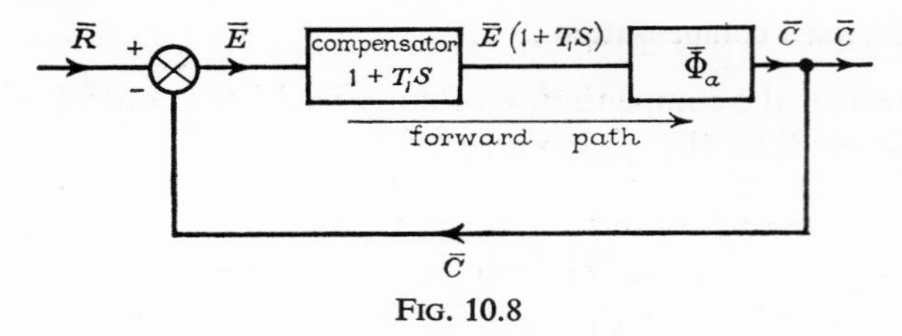

FIG. 10.8

Figure 10.8 shows the compensated system. As the compensation employs the derivative of the error it is called *error derivative compensation.*

Error derivative compensation can be usefully approximated by a passive electric network.

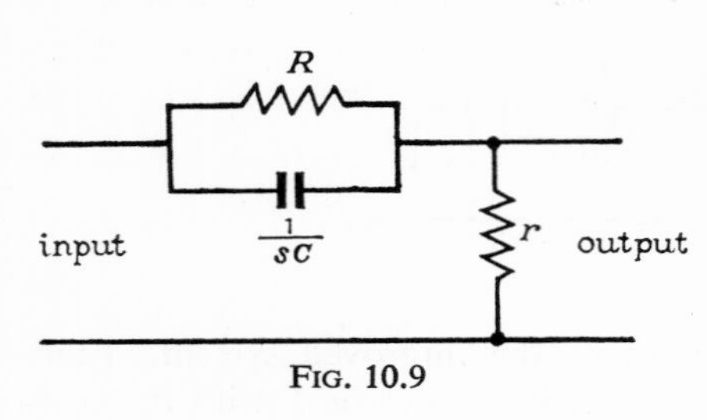

FIG. 10.9

Figure 10.9 shows the passive compensation network. Its transfer function is

$$\bar{\Phi}_c = \frac{r}{R+r} \cdot \frac{(1+sCR)}{(1+sCR(r/R+r))}$$

$$\bar{\Phi}_c = \frac{r}{r + \dfrac{R/sC}{R+1/sC}}$$

Therefore, if $CR = T_1$ and $r/(R+r) = A_1$

$$\bar{\Phi}_c = A_1 \left(\frac{1+sT_1}{1+sT_1A_1} \right) \tag{10.19}$$

Let us consider the compensation network in the forward path of the system.

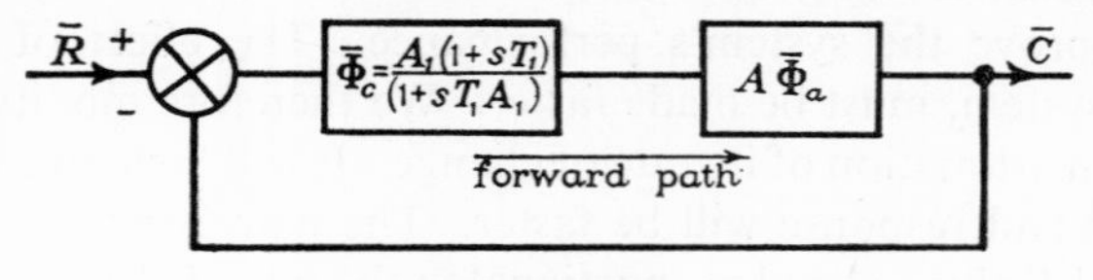

FIG. 10.10

Figure 10.10 shows the compensated system. If A_1 is chosen as less than 0·1 it usefully approximates the system in figure 10.8. If $A_1 < 0{\cdot}1$ the numerator of Φ_c is more effective than the denominator and the compensator transfer function approximates $A_1(1+sT_1)$. The amplifier of the system is readjusted to compensate for the network attenuation. The effective compensation is approximately $(1+sT_1)$. This is called error derivative compensation.

Error integral compensation

Let us consider a system which has a suitable transient response, but has too large a steady-state error. An example of a control system which has a large steady-state velocity error is the remote position control system damped with viscous friction (chapter 4, section 4.1). The viscous friction damping is introduced to give a better degree of stability. It, however, does provide a retarding force proportional to the system output speed. This retardation produces the undesirable steady-state velocity error.

Consider a ramp input to the system. In the steady state the output is a ramp lagging behind the input. The system error is a constant velocity error. If the error is modified by adding a signal proportional to its integral, the original, steady-state error increases with time and the output, activated by this new error, accelerates. The output catches up with the input, hence decreasing the error plus its proportional integral signal. Effectively, in the viscous friction damped remote position control system, the error plus integral of error signal provides an additional motor signal, demanding a greater output torque to accelerate the output shaft (output torque is proportional to the error). The modified error can be expressed as

$$\underset{\text{(original error)}}{\bar{E}} + \underset{\text{(signal proportional to the integral of the error)}}{\frac{\bar{E}}{T_2 s}} = \text{modified error}$$

therefore

$$\text{modified error} = \bar{E}(1+1/sT_2) \tag{10.20}$$

where T_2 is a constant. Equation 10.20 indicates that the transfer function $\bar{\Phi}_c$ of the compensator should be $(sT_2+1)/sT_2$.

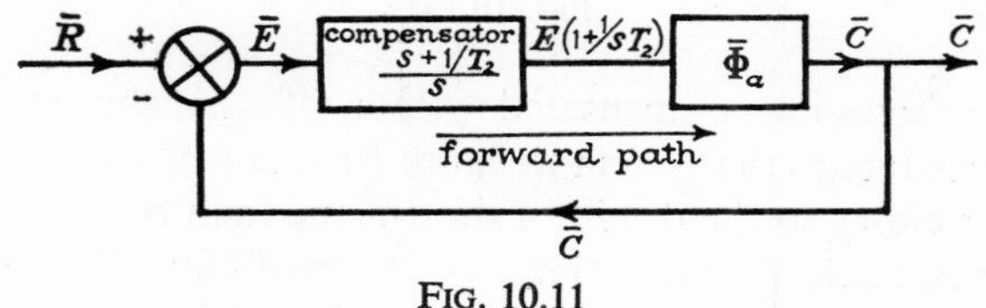

FIG. 10.11

Figure 10.11 shows the compensated system. As the compensation employs the integral of the error it is called *error integral compensation*.

We can consider improvement by integral compensation in terms of system types. A type 1 system has a steady-state velocity error. A type 2 system, however, has a zero velocity error and therefore an increase in system type will remove a velocity error from a type 1 system. The integral compensation $(s+1/T_2)/s$ introduces an extra pole into the open-loop transfer function, thus increasing the system type and removing the velocity error. We have suggested that the uncompensated system transient response is satisfactory. The zero $(s = -1/T_2)$ therefore must be chosen such that the transient response is not altered very much, i.e. the compensated system characteristic equation must remain the same as that for the uncompensated system.

Error integral compensation can be usefully approximated by a passive electric circuit.

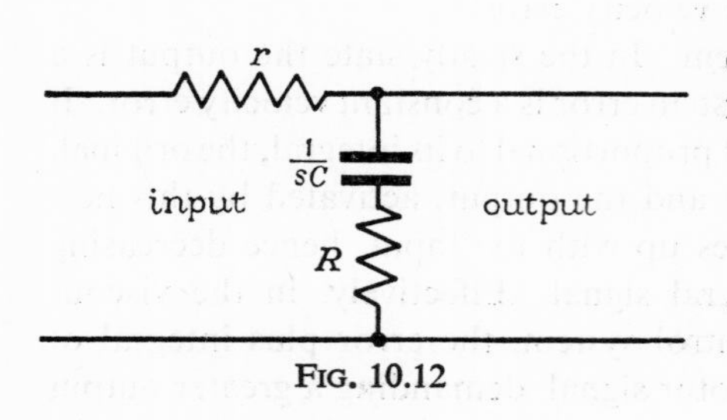

FIG. 10.12

Figure 10.12 shows the passive compensation network. Its transfer function is

$$\bar{\Phi}_c = \frac{R+1/sC}{R+r+1/sC}$$

$$\bar{\Phi}_c = \frac{1+sCR}{1+sCR\left(\frac{R+r}{R}\right)}$$

Therefore if $CR = T_2$ and $R/(R+r) = A_2$

$$\bar{\Phi}_c = A_2\left(\frac{s+1/T_2}{s+A_2/T_2}\right) \quad (10.21)$$

Let us consider the compensation network (transfer function equation 10.21) in the forward path of the system.

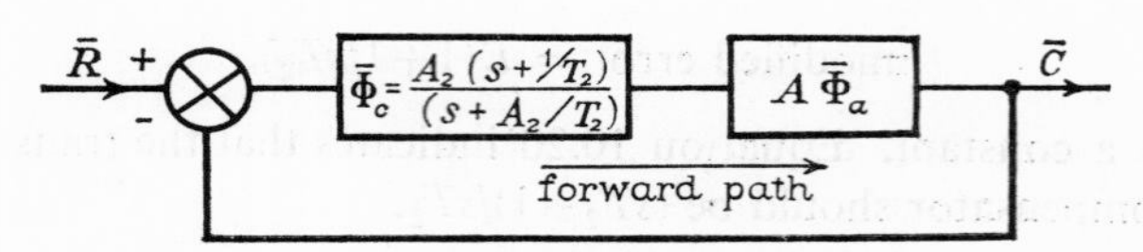

FIG. 10.13

Figure 10.13 shows the compensated system. If A_2 is chosen as less than 1 it usefully approximates the system in figure 10.11, i.e. the pole of the passive compensator is closer to the origin of the s-plane than the zero.

Integral compensation in general makes the possibility of system instability more likely by the introduction of poles near the origin of the s-plane. Consideration of the choice of A_2 and T_2 is given in the following examples.

Error derivative and integral compensation

If a system has both a poor transient response and a large steady-state error, derivative and integral compensation may be employed together to improve the system performance.

The passive networks are usually called *lead compensation* and *lag compensation* networks respectively. This will become clear when we consider an example of compensation with reference to frequency response analysis methods.

10.4 Examples

We shall now consider examples of compensation with reference to the methods of control system analysis discussed in chapters 7 and 8.

Lead compensation: root-locus patterns

It is required to improve the transient response of a control system whose open-loop transfer function is given by

$$\overline{\Phi}_a = \frac{K}{s(s+2)(s+3)}$$

The construction of the root-locus pattern for this system was considered in detail in chapter 7 (figure 7.21). Let the required effective damping ratio be $\zeta = 0{\cdot}53$. The dominant roots are the conjugate complex poles at P_1 and P_2.

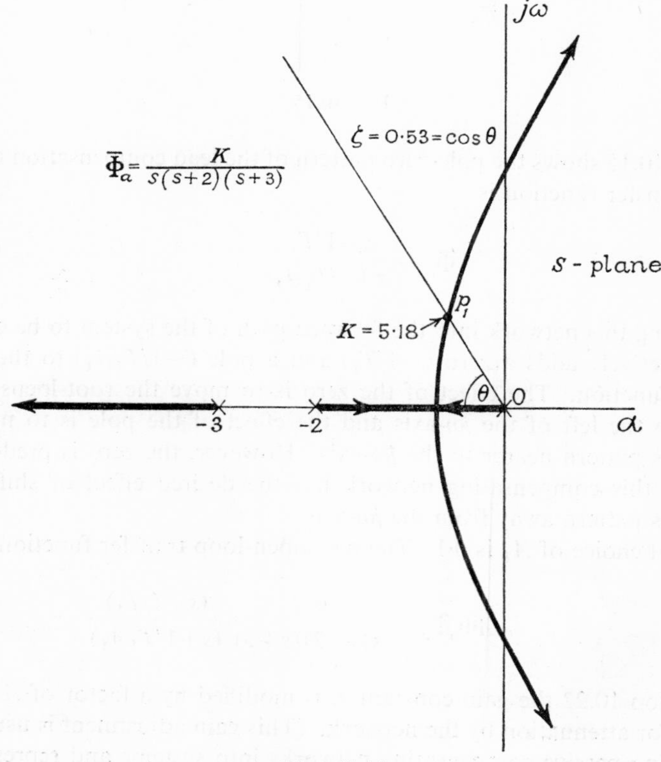

FIG. 10.14

Figure 10.14 shows the root-locus pattern of the system. By comparison with a second-order system (chapter 7, section 4) length OP_1 gives the undamped natural frequency ω_n for a specific effective damping ratio ζ. Therefore, for $\zeta = 0{\cdot}53$, $\omega_n = 1{\cdot}18$ radian per second.

In order to improve the system transient response the undamped natural frequency ω_n must be larger for the same effective damping ratio ζ. This means we must move the root-locus pattern further to the left of the axis (i.e. to make OP_1 longer).

Let us now consider the effect of introducing the passive lead network from its pole-zero pattern.

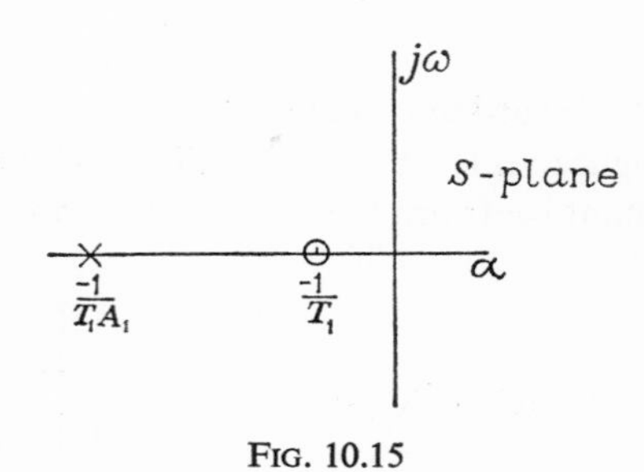

FIG. 10.15

Figure 10.15 shows the pole-zero pattern of the lead compensation network whose transfer function is

$$\bar{\Phi}_c = \frac{s+1/T_1}{s+1/T_1A_1}$$

Introducing this network into the forward path of the system to be compensated effectively adds a zero $(-1/T_1)$ and a pole $(-1/T_1A_1)$ to the system transfer function. The effect of the zero is to move the root-locus pattern further to the left of the $j\omega$-axis and the effect of the pole is to move the root-locus pattern nearer to the $j\omega$-axis. However, the zero is predominant and thus this compensating network has the desired effect of shifting the root-locus pattern away from the $j\omega$-axis.

A usual choice of A_1 is 0·1. The new open-loop transfer function is

$$A\bar{\Phi}_a\bar{\Phi}_c = \frac{KA}{s(s+2)(s+3)} \frac{(s+1/T_1)}{(s+1/T_1A_1)} \tag{10.22}$$

In equation 10.22 the gain constant K is modified by a factor of A to compensate for attenuation by the network. (This gain adjustment is usual when introducing passive compensating networks into systems and represents the modification due to the introduction of elements into a system.)

It is now necessary to choose a value for T_1. This choice can be a matter of trial and error. However, it is reasonable to eliminate a large time constant in the denominator of the uncompensated open-loop transfer function. Let us choose $T_1 = 0{\cdot}5$. (We could have chosen $T_1 = 0{\cdot}33$.) Equation 10.22 now becomes

$$A\bar{\Phi}_a\bar{\Phi}_c = \frac{KA}{s(s+2)(s+3)} \cdot \frac{(s+2)}{(s+20)}$$

$$A\bar{\Phi}_a\bar{\Phi}_c = \frac{KA}{s(s+3)(s+20)} \qquad (10.23)$$

Equation 10.23 is the compensated system open-loop transfer function.

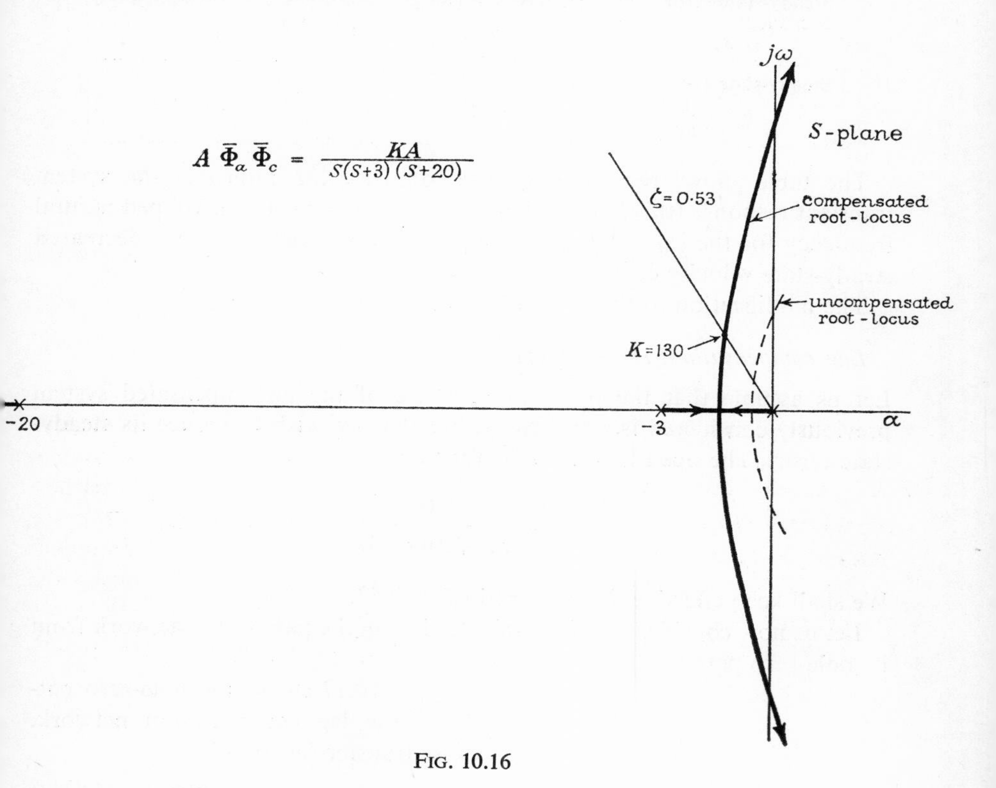

FIG. 10.16

Figure 10.16. We shall now compare quantities taken from the root-locus pattern of the compensated system, as shown in this figure, with those from the uncompensated system. We shall also evaluate and compare steady-state errors. For direct comparison this is presented in a tabular form.

Quantity	Uncompensated system $\Phi_a = \dfrac{5\cdot18}{s(s+2)(s+3)}$	Compensated system passive lead compensation $A\Phi_a\Phi_c = \dfrac{5\cdot18A}{s(s+3)(s+20)}$
Effective damping ratio ζ	0·53	0·53
Gain constant K	5·18	$5\cdot18A = 130\cdot16$
Undamped natural frequency ω_n (length OP_1)	1·18 radian per second	2·54 radian per second
Steady-state error coefficient K_v	$[K_v]_{t\to\infty} = [s\Phi_a]_{s\to 0}$ 0·86	$[K_v]_{t\to\infty} = [sA\Phi_a\Phi_c]_{s\to 0}$ 2·17
Steady-state error $E_v = \dfrac{1}{K_v}$	1·16	0·46

The table shows that passive lead compensation improves the system transient response (this is indicated by the increase in the undamped natural frequency for the same effective damping ratio). There is also a decreased steady-state velocity error.

The modification to the system gain constant is $A = 25\cdot13$.

Lag compensation: root-locus patterns

Let us assume that the transient response of the uncompensated system previously considered is satisfactory, but that we wish to reduce its steady-state error. The open-loop transfer function is

$$\overline{\Phi}_a = \frac{5\cdot18}{s(s+2)(s+3)}$$

We shall keep effective damping ratio $\zeta = 0\cdot53$.

Let us now consider the effect of introducing the passive lag network from its pole-zero pattern.

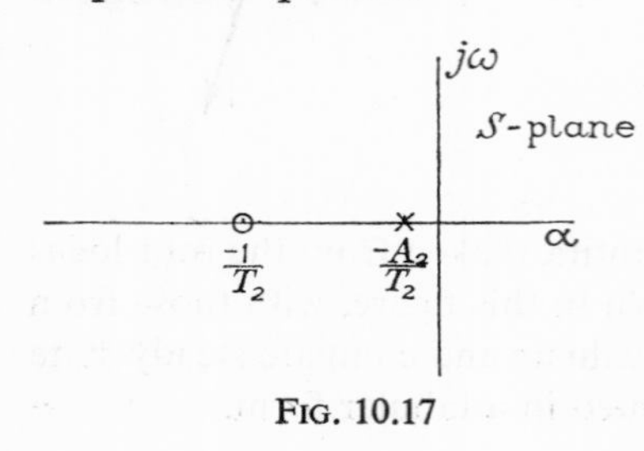

FIG. 10.17

Figure 10.17 shows the pole-zero pattern of the lag compensation network, whose transfer function is

$$\overline{\Phi}_c = A_2\left(\frac{s+1/T_2}{s+A_2/T_2}\right)$$

The open-loop transfer function of the compensated system is

$$A\bar{\Phi}_c\bar{\Phi}_a = \frac{5{\cdot}18A}{s(s+2)(s+3)} \cdot A_2\left(\frac{s+1/T_2}{s+A_2/T_2}\right) \tag{10.24}$$

To reduce the steady-state velocity error the uncompensated type 1 system (equation 10.23) must approximate to a type 2 system. This means that equation 10.24 must approach the open-loop transfer function of a type 2 system. Therefore, with reference to the uncompensated system pole-zero pattern, the pole ($s = -A_2/T_2$) is made comparatively very close to the s-plane origin.

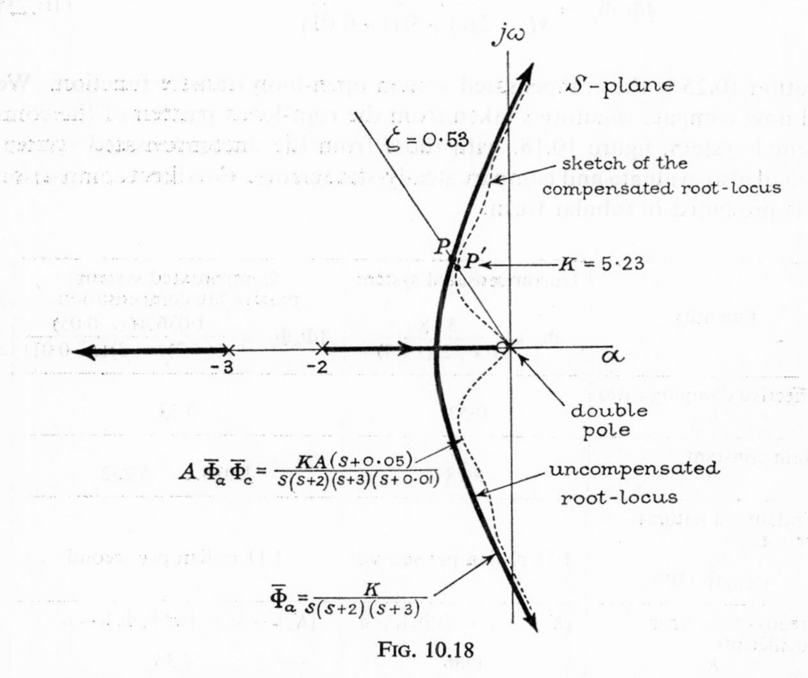

FIG. 10.18

Figure 10.18. Let the position of the pole be $s = -0{\cdot}01$. In choosing the position of the zero of the compensator we must remember that the transient response of the system is to be altered as little as possible. This means that dominant roots of the compensated and uncompensated systems must be much the same. This is ensured by choosing P', on the s-plane, along the $\zeta = 0{\cdot}53$ damping-ratio line near P_1.

We can find the compensator zero from equation 7.14. Its position is adjusted along the negative real axis, with reference to the point P' on the root-locus. In the example the point P' is near enough to P_1 to ensure that the dominant roots of the closed-loop system characteristic equation are very nearly the same. The zero was found to be at $s = -0{\cdot}05$, which gives

$T_2 = 20$ and $A_2 = 0{\cdot}2$. It can be seen that the pole and zero of the compensator are relatively close together and this is characteristic of a passive lag network compensator. The compensated system open-loop transfer function is

$$A\overline{\Phi}_a\overline{\Phi}_c = \frac{5{\cdot}18\,A}{s(s+2)(s+3)}\,.\,0{\cdot}2\left(\frac{s+0{\cdot}05}{s+0{\cdot}01}\right)$$

$$A\overline{\Phi}_a\overline{\Phi}_c = \frac{1{\cdot}036\,A(s+0{\cdot}05)}{s(s+2)(s+3)(s+0{\cdot}01)} \tag{10.25}$$

Equation 10.25 is the compensated system open-loop transfer function. We shall now compare quantities taken from the root-locus pattern of the compensated system, figure 10.18, with those from the uncompensated system. We shall also evaluate and compare steady-state errors. For direct comparison this is presented in tabular form.

Quantity	Uncompensated system $\Phi_a = \frac{5{\cdot}18}{s(s+2)(s+3)}$	Compensated system passive lag compensation $A\Phi_a\Phi_c = \frac{1{\cdot}036A(s+0{\cdot}05)}{s(s+2)(s+3)(s+0{\cdot}01)}$
Effective damping ratio ζ	0·53	0·53
Gain constant K	5·18	$1{\cdot}036A = 5{\cdot}232$
Undamped natural frequency ω_n (length OP′)	1·18 radian per second	1·11 radian per second
Steady-state error coefficients K_v	$[K_v]_{t \to \infty} = [s\Phi_a]_{s \to 0}$ 0·86	$[K_v]_{t \to \infty} = [sA\Phi_a\Phi_c]_{s \to 0}$ 4·36
Steady-state error $E_v = \frac{1}{K_v}$	1·16	0·23

The table shows that passive lag compensation reduces the system steady-state error considerably. There is a slight deterioration on the system transient response which is indicated by the reduction of the undamped natural frequency.

The modification to the system gain constant is $A = 5{\cdot}05$.

By keeping the effective damping ratio the same, approximately the same degree of stability is maintained.

Comments

The increase in the undamped natural frequency ω_n, by lead compensation, results in a faster response, a faster settling time and a reduction in the steady-state error.

The reduction of the steady-state error E_v (i.e. comparatively greater gain for the elements contributing to the lagging response only) by lag compensation gives a slight deterioration of the system transient response, therefore settling time is increased.

Compensation: Nyquist diagrams

The dominant conjugate complex roots of the control systems we are considering make it justifiable for a comparison with second-order system behaviour. Even though the system may be third order and above it is helpful to consider an undamped natural frequency ω_n and an effective damping ratio ζ. (This was done for root-locus patterns in chapter 7.) It is necessary to obtain a comparison of the second-order ω_n and ζ with reference to the frequency response loci.

Consider the general second-order system (remote position control system chapter 4, section 1) whose open-loop transfer function is

$$\bar{\Phi}_a = \frac{\omega_n^2}{s(s+2\zeta\omega_n)}$$

In terms of frequency the transfer function becomes

$$\Phi_a(j\omega) = \frac{\omega_n^2/2\zeta\omega_n}{j\omega(j\omega/2\zeta\omega_n+1)}$$

The closed-loop transfer function is

$$\Phi(j\omega) = \frac{\Phi_a(j\omega)}{1+\Phi_a(j\omega)} = \frac{\omega_n^2}{-\omega^2+\omega_n^2+j2\zeta\omega\omega_n} \tag{10.26}$$

The magnitude M at any frequency is given by the modulus of equation 10.26.

$$M = |\Phi(j\omega)| = \frac{\omega_n^2}{\sqrt{[(\omega_n^2-\omega^2)^2+4\zeta^2\omega^2\omega_n^2]}} \tag{10.27}$$

The frequency at which the maximum magnitude $M_{\max}$ occurs is found by differentiating equation 10.27 and equating to zero.

$$\frac{dM}{d\omega} = 0 = \frac{\omega_n^2[-\frac{1}{2}(4\omega^3-4\omega_n^2\omega+8\zeta^2\omega_n^2\omega)]}{[(\omega_n^2-\omega^2)^2+4\zeta^2\omega^2\omega_n^2]^{\frac{3}{2}}}$$

therefore

$$0 = \omega_n^2-\omega^2-2\omega_n^2\zeta^2 \tag{10.28}$$

The frequency at which M_{max} occurs is the system resonant frequency $\omega = \omega_r$ and from equation 10.28

$$\omega_r = \omega_n\sqrt{(1-2\zeta^2)} \tag{10.29}$$

On a calibrated Nyquist diagram the M circle tangential to the frequency response locus gives M_{max}. The resonant frequency, for a specific system damping ratio ζ, is proportional to the undamped natural frequency ω_n and hence gives a guide to the system transient response. Thus an increase of ω_r means a faster system response. Equation 10.29 demonstrates this. However, as can be seen from the equation, ω_r is only proportional to ω_n for damping ratios less than $\zeta = 0{\cdot}707$. For values of damping ratio greater than 0·707 the left-hand side of the equation is imaginary. M_{max} is found by substituting the resonant frequency, equation 10.29, into equation 10.27. Thus we obtain

$$M_{max} = \frac{1}{2\zeta\sqrt{(1-\zeta^2)}} \tag{10.30}$$

Equation 10.30 indicates that for a specific damping ratio ζ, there is a particular M_{max}. Hence, reading the value of M_{max} from a Nyquist diagram, we can obtain the damping ratio ζ (for example $M_{max} = 1{\cdot}113$ gives $\zeta = 0{\cdot}53$). We can say that systems with dominant conjugate complex roots have an *effective* damping ratio ζ, which can be estimated in terms of the maximum magnitude M_{max} (i.e. M_{max} indicates the degree of stability of a system).

In considering compensation with reference to Nyquist diagrams (and Nichols diagrams) we shall specify the system performance in terms of maximum magnitude M_{max} (degree of stability), resonant frequency ω_r (transient response) and the steady-state error. The steady-state errors are obtained from the open-loop transfer function using the Laplace final-value theorem, but with the complex variable s replaced by $j\omega$.

In terms of frequency response the general open-loop transfer function can be expressed

$$\Phi_a(j\omega) = \frac{K(j\omega+a_1)(j\omega+a_2)(j\omega+a_3)\ldots(j\omega+a_m)}{j\omega^n(j\omega+b_1)(j\omega+b_2)(j\omega+b_3)\ldots(j\omega+b_p)}$$

$$\Phi_a(j\omega) = \frac{\dfrac{K(a_1\,a_2\,a_3\ldots a_m)}{b_1\,b_2\,b_3\ldots b_p}(j\omega/a_1+1)(j\omega/a_2+1)\ldots(j\omega/a_m+1)}{j\omega^n(j\omega/b_1+1)(j\omega/b_2+1)\ldots(j\omega/b_p+1)} \tag{10.31}$$

Equation 10.31 is the open-loop transfer function rearranged such that the steady-state error coefficient

$$K\frac{(a_1\,a_2\,a_3\ldots a_m)}{(b_1\,b_2\,b_3\ldots b_p)} = K'$$

appears in the numerator. In the systems we shall consider the steady-state error coefficient is the velocity coefficient K_v.

Lead compensation: Nyquist diagrams

(For comparison we shall consider the same system for compensation that was considered with reference to root-locus patterns.)

It is required to improve the transient response of a control system whose open-loop transfer function is given by

$$\Phi_a(j\omega) = \frac{K}{j\omega(j\omega+2)(j\omega+3)}$$

$$\Phi_a(j\omega) = \frac{K'}{j\omega(j\omega/2+1)(j\omega/3+1)} \tag{10.32}$$

The effective damping ratio ζ is 0·53, i.e. $M_{max} = 1{\cdot}113$. The resonant frequency ω_r is 0·76 radian per second. (M_{max} was obtained from equation 10.30).

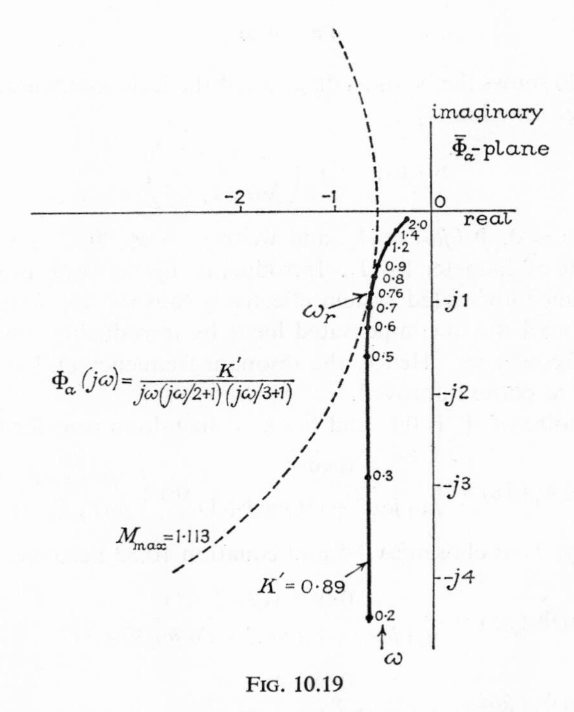

FIG. 10.19

Figure 10.19 shows the Nyquist diagram for the uncompensated system described by the open-loop transfer function equation 10.32. The locus has been adjusted for gain by the construction shown in chapter 8, section 3 and K' is found to be 0·89 (page 150).

In order to improve the system transient response it is necessary to increase the resonant frequency ω_r while keeping the value of maximum magnitude M_{max} the same. The Nyquist locus must be reshaped so that the frequencies are 'pushed' anti-clockwise around the locus, i.e. ω_r becomes larger.

With the help of its Nyquist locus, let us now consider the effect of introducing the passive lead network.

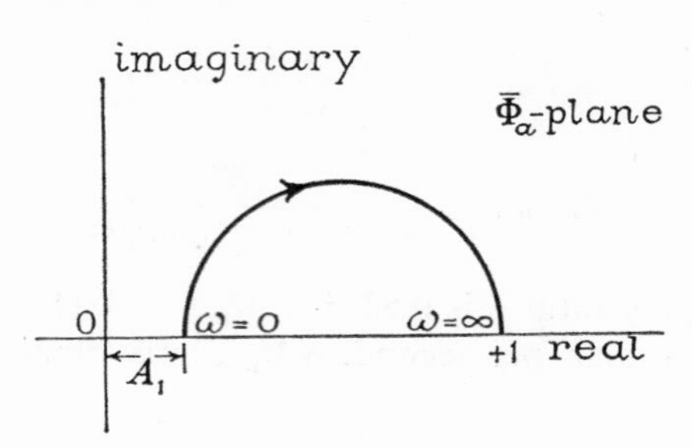

FIG. 10.20

Figure 10.20 shows the Nyquist diagram of the lead compensation network whose transfer function is

$$\Phi_c(j\omega) = A_1\left(\frac{j\omega T_1+1}{j\omega T_1 A_1+1}\right)$$

Since when $\omega = 0$, $\Phi_c(j\omega) = A_1$, and when $\omega = \infty$, $\Phi_c(j\omega) = 1$, the locus is a semi-circle of diameter $1-A_1$. Introducing this network into the forward path of the uncompensated system effectively 'pushes' the frequencies anti-clockwise around the uncompensated locus by introducing a positive phase angle at all frequencies. Hence, the resonant frequency ω_r is increased and the transient response improved.

A usual choice of A_1 is 0·1, and the new open-loop transfer function is

$$A\Phi_a(j\omega)\,\Phi_c(j\omega) = \frac{0{\cdot}89\,A}{j\omega(j\omega/2+1)(j\omega/3+1)}\,0{\cdot}1\left(\frac{j\omega T_1+1}{j\omega T_1 A_1+1}\right) \qquad (10.33)$$

As previously, T_1 is chosen as 0·5 and equation 10.33 becomes

$$A\Phi_a(j\omega)\,\Phi_c(j\omega) = \frac{0{\cdot}089\,A(j\omega/2+1)}{j\omega(j\omega/2+1)(j\omega/3+1)(j\omega/20+1)}$$

$$A\Phi_a(j\omega)\,\Phi_c(j\omega) = \frac{0{\cdot}089\,A}{j\omega(j\omega/3+1)(j\omega/20+1)} \qquad (10.34)$$

Equation 10.34 is the compensated system open-loop transfer function.

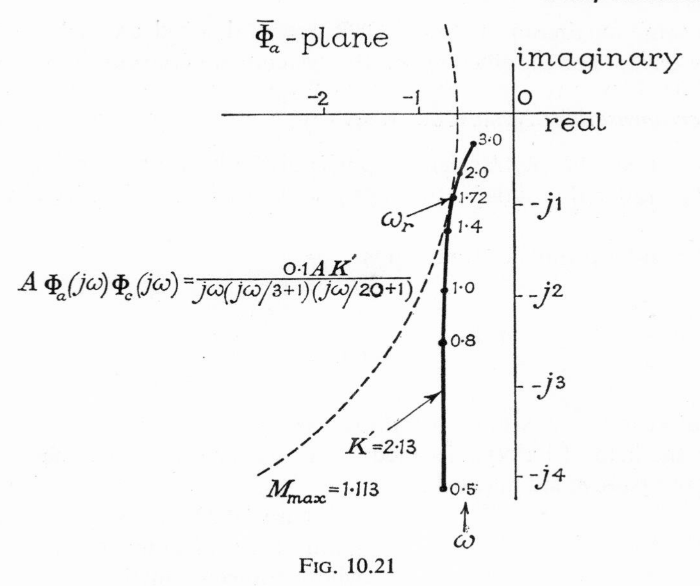

FIG. 10.21

Figure 10.21 shows the Nyquist diagram of the compensated system. We shall now compare quantities taken from it with those taken from the uncompensated system. We shall also evaluate and compare steady-state errors. For direct comparison this is presented in a tabular form.

Quantity	Uncompensated system $\Phi_a(j\omega) = \dfrac{K'}{j\omega(j\omega/2+1)(j\omega/3+1)}$	Compensated system passive lead compensation $A\Phi_a(j\omega)\Phi_c(j\omega) = \dfrac{0{\cdot}1\,AK'}{j\omega(j\omega/3+1)(j\omega/20+1)}$
Maximum magnitude M_{max}	1·113 (effective ζ = 0·53)	1·113 (effective ζ = 0·53)
Resonant frequency ω_r	0·76 radian per second	1·72 radian per second
Steady-state error coefficient K_v	$K' = 0{\cdot}89$	$0{\cdot}1\,K'A = 2{\cdot}13$
Steady-state error $E_v = \dfrac{1}{K_v}$	1·124	0·47

The table shows that passive lead compensation improves the system transient performance. (This is indicated by the increase in resonant frequency for the same maximum magnitude.) There is also a decreased steady-state velocity error. The modification to the system gain constant is $A = 24$.

Lag compensation: Nyquist diagrams

Let us assume that the transient response of the uncompensated system previously considered is satisfactory, but that we wish to reduce its steady-state error.

The open-loop transfer function is

$$\Phi_a(j\omega) = \frac{0{\cdot}89}{j\omega(j\omega/2+1)(j\omega/3+1)} \qquad (10.32)$$

We shall keep the maximum magnitude $M_{max} = 1{\cdot}113$.

With the help of its Nyquist locus let us now consider the effect of introducing the passive lag network.

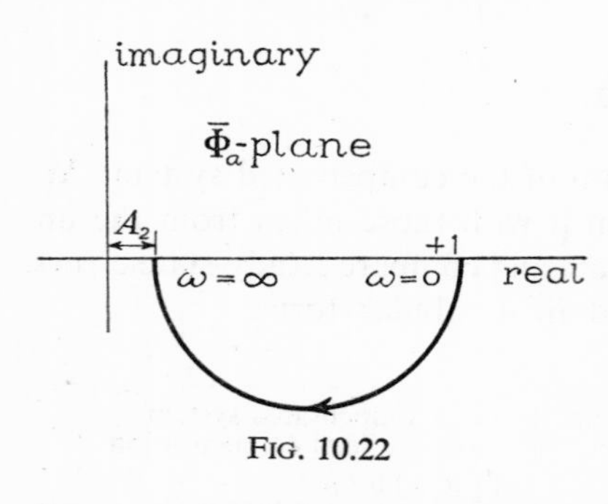

FIG. 10.22

Figure 10.22 shows the Nyquist diagram of the lag compensation network, whose transfer function is

$$\Phi_c(j\omega) = \frac{1+j\omega T_2}{1+j\omega T_2/A_2} \qquad (10.35)$$

The locus is a semi-circle of diameter $1-A_2$ (when $\omega=0$, $\Phi(j\omega)=1$, and when $\omega=\infty$, $\Phi_c(j\omega)=A_2$). Introducing this network into the forward path of a system enables the steady-state velocity error, due to poor response at the lower frequencies, to be decreased. In order to do this the compensator attenuates at high frequencies, but not over a range of low frequencies. The system gain can then be increased such that the gain is increased comparatively more over the low-frequency range than at the high frequencies. (High and low frequencies refer to the upper and lower ends of the appreciable bandwidth of the system and are entirely comparative terms.) The steady-state velocity error due to insufficient gain at low frequencies is removed by the compensator and an overall gain adjustment.

As previously, let $A_2 = 0{\cdot}2$. It is now necessary to choose a value for T_2. The transient response of the system is satisfactory and should be altered as little as possible. This means that the resonant frequency $\omega_r = 0{\cdot}76$ radian per second, or a frequency close to it, must still occur at the specified maximum magnitude $M_{max} = 1{\cdot}113$. Therefore the introduction of the compensator should produce as small a phase shift of the calibrated position of ω_r as possible. The amount of phase shift that can be tolerated depends entirely on how much deterioration of the transient response can be tolerated by the designer.

We shall consider a phase shift of $-3°$ (the passive lag network introduces a phase lag at all frequencies). For this specified phase shift the position of ω_r ($=0{\cdot}76$ radian per second) on the Nyquist locus of the compensator will be considered to be at a phase angle of $-3°$.

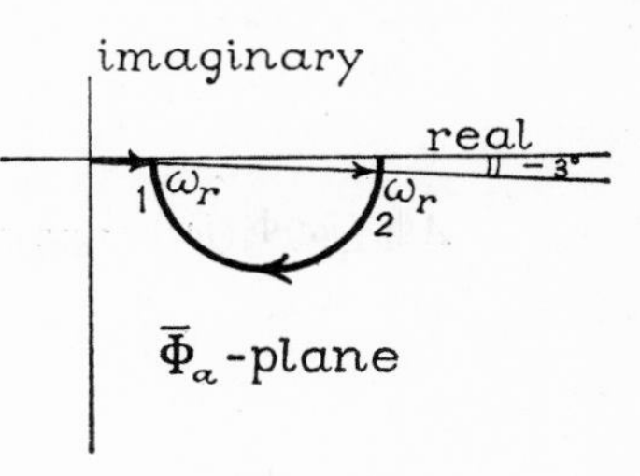

FIG. 10.23

Figure 10.23. This figure shows the two possible positions of ω_r on the compensator Nyquist locus. Substituting $\omega_r=0{\cdot}76$ radian per second and $A_2=0{\cdot}2$ into equation 10.35, we have

$$\Phi_c(j\omega) = \frac{1+j0{\cdot}76T_2}{1+j0{\cdot}76T_2/0{\cdot}2}$$

For a phase angle of $-3°$

$$|\,\Phi_c(j\omega)\,|\,\angle -3° = \frac{|\,1+j0{\cdot}76T_2\,|\,\angle \tan^{-1} 0{\cdot}76T_2}{|\,1+j3{\cdot}8\;T_2\,|\,\angle \tan^{-1} 3{\cdot}8\;T_2}$$

To find the value of T_2 we equate phase angles

$$-3° = \tan^{-1}(0{\cdot}76T_2) - \tan^{-1}(3{\cdot}8T_2)$$

therefore,
$$\tan(-3°) = \frac{0{\cdot}76T_2 - 3{\cdot}8T_2}{1+(0{\cdot}76)(3{\cdot}8)T_2^2}$$

Solving, we have $T_2 \simeq 0$, or $T_2 = 20{\cdot}2$

For a larger attenuation over the 'high' frequency range the value $T_2 = 20{\cdot}2$ is chosen. In order to have a direct comparison with passive lag network compensation, with reference to root-locus patterns, the value of T_2 is taken as 20 (it is usual to round off the figure obtained for T_2).

The compensator transfer function is

$$\Phi_c(j\omega) = \frac{1+j\omega 20}{1+j\omega 100}$$

Equation 10.32 is modified by this compensator, and the open-loop transfer function of the compensated system is

$$A\Phi_a(j\omega)\,\Phi_c(j\omega) = \frac{0{\cdot}89\,A(1+j\omega 20)}{j\omega(j\omega/2+1)(j\omega/3+1)(1+j\omega 100)}$$

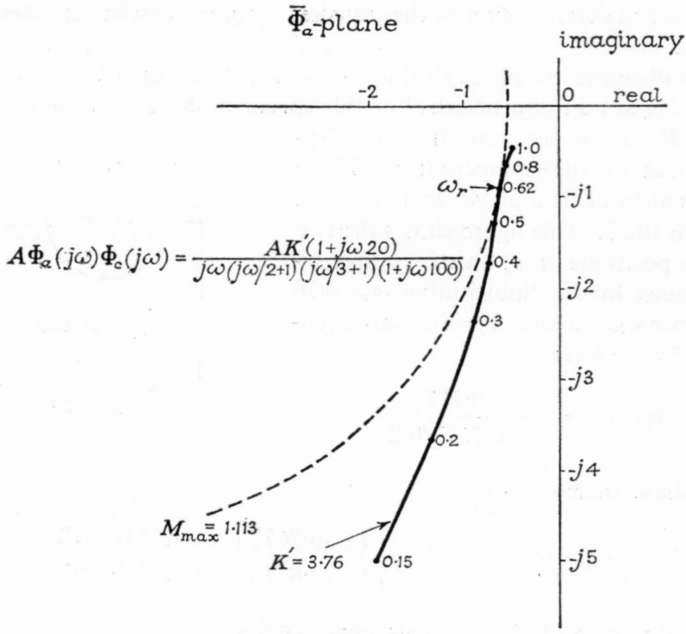

FIG. 10.24

Figure 10.24 shows the Nyquist diagram of the compensated system. We shall now compare quantities taken from the Nyquist diagram of the compensated system with those from the uncompensated system. We shall also evaluate and compare steady-state errors. For direct comparison this is presented in tabular form.

Quantity	Uncompensated system $\Phi_a(j\omega) = \frac{K'}{j\omega(j\omega/2+1)(j\omega/3+1)}$	Compensated system passive lag compensation $A\Phi_a(j\omega)\Phi_c(j\omega) = \frac{K'A(1+j\omega 20)}{j\omega(j\omega/2+1)(j\omega/3+1)(1+j\omega 100)}$
Maximum magnitude M_{max}	1·113 (effective $\zeta = 0{\cdot}53$)	1·113 (effective $\zeta = 0{\cdot}53$)
Resonant frequency ω_r	0·76 radian per second	0·62 radian per second
Steady-state error coefficient K_v	$K_v = K' = 0{\cdot}89$	$K_v = K'A = 3{\cdot}76$
Steady-state error $E_v = \frac{1}{K_v}$	1·124	0·266

The table shows that passive lag compensation reduces the system steady-state error considerably. There is a deterioration in the system transient response which is indicated by the reduction of the resonant frequency.

The modification to the system gain constant is $A = 4{\cdot}22$.

Compensation: Bode diagrams

We shall consider system performance in terms of comparison with a second-order system with reference to Bode diagrams. But first let us examine Bode diagrams in order to obtain the steady-state error coefficient K'. The general open-loop transfer function, equation 10.31, can be rewritten as

$$\Phi_a(j\omega) = \frac{K'N(j\omega)}{(j\omega)^n d(j\omega)} \tag{10.36}$$

where $N(j\omega) = (j\omega/a_1+1)(j\omega/a_2+1)(j\omega/a_3+1)\ldots(j\omega/a_m+1)$

$d(j\omega) = (j\omega/b_1+1)(j\omega/b_2+1)(j\omega/b_3+1)\ldots(j\omega/b_p+1)$

and $n = 0, 1, 2, 3$, etc., defines the system type. The gain curve (logarithm of the magnitude in decibels against the logarithm of frequency) is given by

$$\begin{aligned} \text{gain} &= 20\log_{10}|\Phi_a(j\omega)| \\ &= 20\log_{10}K' + 20\log_{10}\left|\frac{N(j\omega)}{d(j\omega)}\right| - n20\log_{10}|j\omega| \end{aligned} \tag{10.37}$$

As the frequency ω tends to zero equation 10.37 becomes

$$[\text{gain}]_{\omega\to 0} = 20\log_{10}K' - n20\log_{10}|j\omega| \tag{10.38}$$

Equation 10.38 represents the start of the gain curve. Let us consider this equation for various values of n:

if $n = 0$, equation 10.38 is a constant,
if $n = 1$, equation 10.38 is a straight line of slope -20dB per decade,
if $n = 2$, equation 10.38 is straight line of slope -40dB per decade, etc.

Therefore, the initial slope of the gain curve indicates the system type and hence the type of steady-state error. The magnitude of the steady-state error can be found by using the Laplace final-value theorem. Let us consider the general second-order system, whose open-loop transfer function is given by

$$\bar{\Phi}_a = \frac{\omega_n^2}{s(s+2\zeta\omega_n)}$$

$$\Phi_a(j\omega) = \frac{\omega_n^2/2\zeta\omega_n}{j\omega(j\omega/2\zeta\omega_n+1)} \tag{10.39}$$

This is a type 1 system, whose steady-state velocity error coefficient is $K_v = \omega_n/2\zeta$.

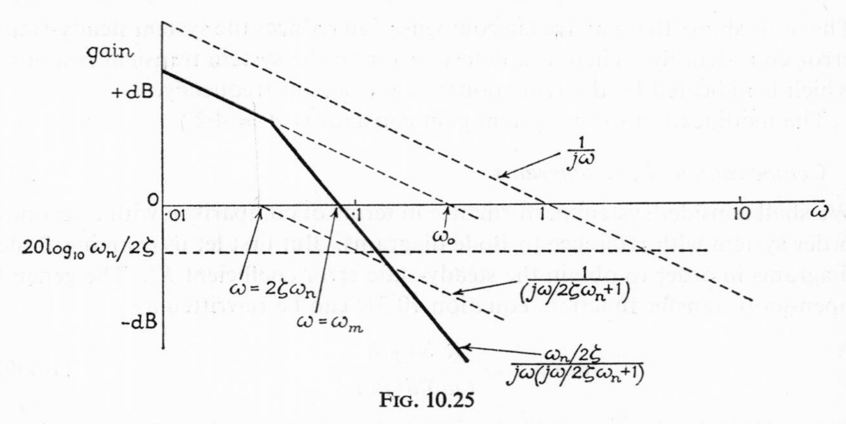

FIG. 10.25

Figure 10.25 shows the gain curve of the second-order system. ω_0 occurs when the gain of equation 10.38 (i.e. the equation of the line representing the initial slope) is zero.

$$0 = 20\log_{10} K' - n20\log_{10}|j\omega_0|$$

in this case $n = 1$ and $K' = K_v$, therefore

$$K_v = \omega_0 \tag{10.40}$$

Phase margin is defined as the amount which the phase falls short of 180° when the magnitude of $\Phi_a(j\omega)$ is unity. The frequency $\omega = \omega_m$ at which the phase margin θ_m occurs is found from equation 10.39 and the definition of θ_m. Therefore

$$1 = \left|\frac{\omega_n^2}{j\omega_m(j\omega_m + 2\zeta\omega_n)}\right|$$

therefore
$$\omega_m = \omega_n[-2\zeta^2 + \sqrt{(4\zeta^4+1)}]^{\frac{1}{2}} \tag{10.41}$$

The phase margin from equation 10.39 is given by

$$\theta_m = \tan^{-1} 2\zeta\omega_n/\omega_m \tag{10.42}$$

substituting equation 10.41 into 10.42 we have

$$\theta_m = \tan^{-1}\left(\frac{4\zeta^2}{-2\zeta^2 + \sqrt{(4\zeta^4+1)}}\right)^{\frac{1}{2}} \tag{10.43}$$

Equation 10.41 gives a relationship between the phase margin frequency ω_m and the undamped natural frequency ω_n for a particular damping ratio ζ. Equation 10.43 gives a particular value of phase margin θ_m for a particular damping ratio ζ. We can say that systems with dominant conjugate complex roots have an *effective* damping ratio which can be estimated in terms of the phase margin θ_m.

In considering compensation, with reference to Bode diagrams, we shall specify the system performance in terms of phase margin θ_m (degree of stability), phase margin frequency ω_m (transient response), and the steady-state error (this is obtained from the intercept of the initial slope of the gain curve with the frequency axis, equation 10.40).

Lead compensation: Bode diagrams

It is required to improve the transient response of a control system whose open-loop transfer function is given by

$$\Phi_a(j\omega) = \frac{0{\cdot}78}{j\omega(j\omega/2+1)(j\omega/3+1)} \tag{10.44}$$

The effective damping ratio ζ is 0·53, i.e. the phase margin is $\theta_m = 54°$. The phase margin frequency ω_m is 0·78 radian per second. (θ_m was obtained from equation 10.43.)

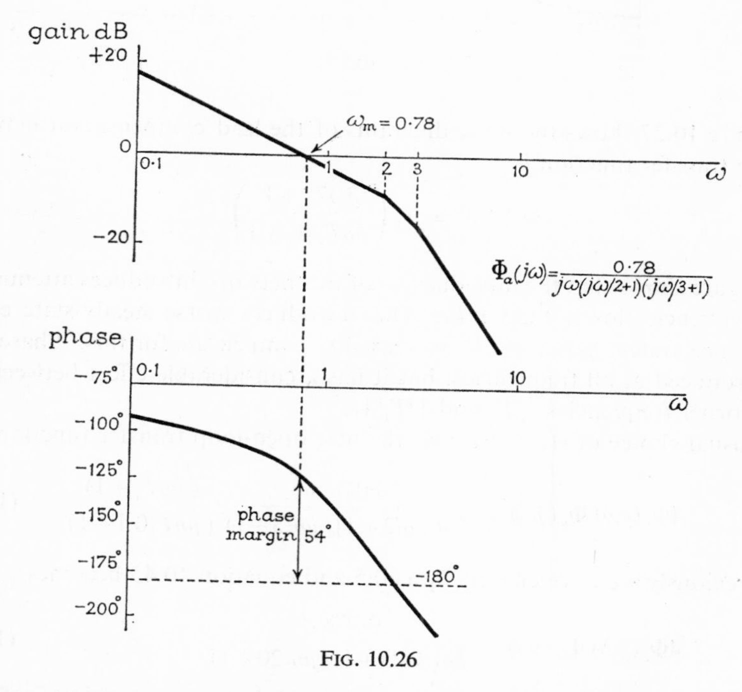

FIG. 10.26

Figure 10.26 shows the Bode diagrams for the uncompensated system described by equation 10.44.

Let us now consider the effect of introducing the passive lead network from its Bode diagrams.

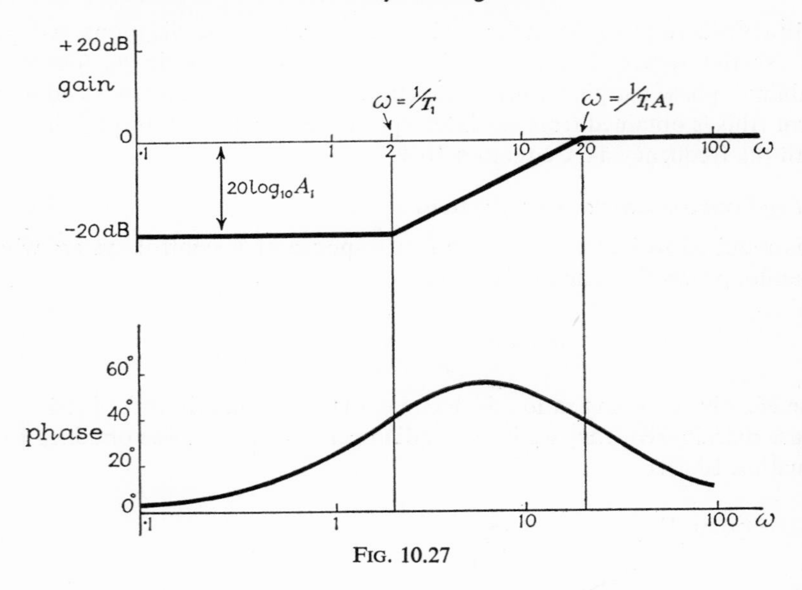

FIG. 10.27

Figure 10.27 shows the Bode diagrams of the lead compensation network whose transfer function is

$$\Phi_c(j\omega) = A_1\left(\frac{j\omega T_1+1}{j\omega T_1 A_1+1}\right)$$

The figure shows that the introduction of this network introduces attenuation for frequencies lower than $1/T_1$. This introduces worse steady-state errors and hence system gain must be increased to compensate for this. Phase lead is introduced at all frequencies, but it has a considerable effect between the two corner frequencies $1/T_1$ and $1/T_1A_1$.

A usual choice of A_1 is 0·1, and the new open-loop transfer function is

$$A\Phi_c(j\omega)\,\Phi_a(j\omega) = \frac{0{\cdot}078\,A}{j\omega(j\omega/2+1)(j\omega/3+1)}\,\frac{(j\omega T_1+1)}{(j\omega T_1 0{\cdot}1+1)} \qquad (10.45)$$

As previously we have chosen T_1 as 0·5 and equation 10.45 becomes

$$A\Phi_c(j\omega)\,\Phi_a(j\omega) = \frac{0{\cdot}078\,A}{j\omega(j\omega/3+1)(j\omega/20+1)} \qquad (10.46)$$

Equation 10.46 is the compensated system open-loop transfer function. To have an improvement of transient response, for a particular phase margin (i.e. an effective damping ratio), we expect a higher value of the phase margin frequency ω_m.

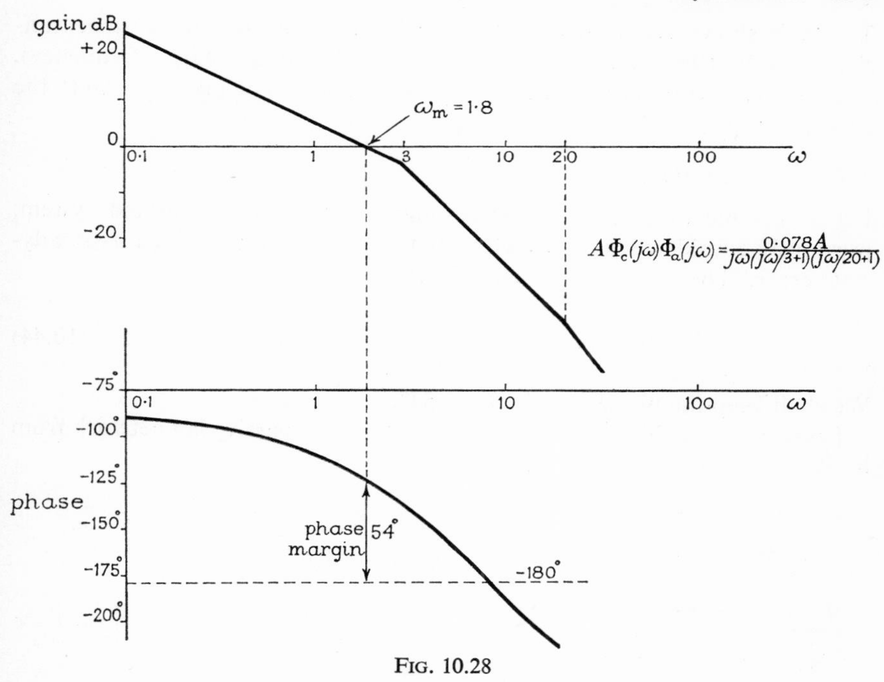

FIG. 10.28

Figure 10.28. We shall now compare quantities taken from the Bode diagrams of the compensated system shown in this figure with those from the uncompensated system. We shall also evaluate and compare steady-state errors. For direct comparison this is presented in tabular form.

Quantity	Uncompensated system $\Phi_a(j\omega) = \frac{0{\cdot}78}{j\omega(j\omega/2+1)(j\omega/3+1)}$	Compensated system passive lead compensation $A\Phi_c(j\omega)\Phi_a(j\omega) = \frac{0{\cdot}078\,A}{j\omega(j\omega/3+1)(j\omega/20+1)}$
Phase margin θ_m	54° (effective $\zeta = 0{\cdot}53$)	54° (effective $\zeta = 0{\cdot}53$)
Phase margin frequency ω_m	0·78 radian per second	1·8 radian per second
Steady-state error coefficient $K' = K_v$	0·78	1·8
Steady-state error $E_v = \frac{1}{K_v}$	1·28	0·56

The table shows that passive lead compensation improves the system transient response (this is indicated by the increase of the phase margin frequency).

There is also a reduction of the steady-state error. The modification to the system gain is $A = 23{\cdot}08$.

Lag compensation: Bode diagrams

Let us assume that the transient response of the uncompensated system, previously considered, is satisfactory but that we wish to reduce its steady-state error. The open-loop transfer function is

$$\Phi_a(j\omega) = \frac{0{\cdot}78}{j\omega(j\omega/2+1)(j\omega/3+1)} \tag{10.44}$$

We shall keep the phase margin $\theta_m = 54°$.

Let us now consider the effect of introducing the passive lag network from its Bode diagrams.

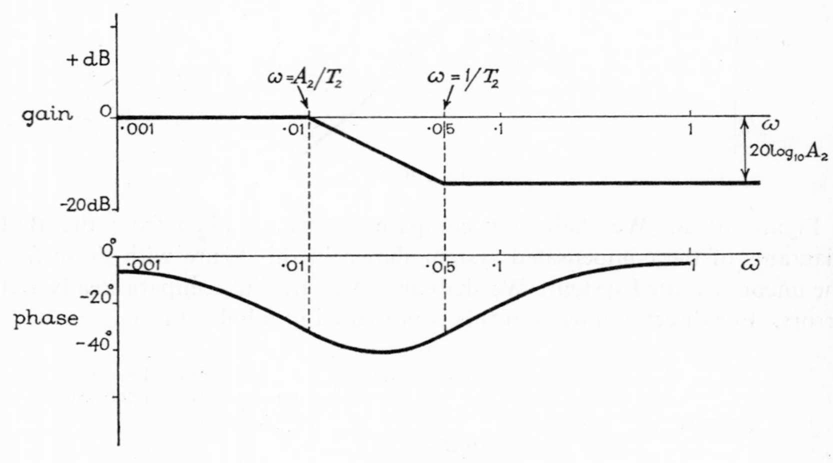

FIG. 10.29

Figure 10.29 shows the Bode diagrams of the lag compensation network whose transfer function is

$$\Phi_c(j\omega) = \frac{1+j\omega T_2}{1+j\omega T_2/A_2}$$

The figure shows that the introduction of this network into the forward path of a system causes attenuation for frequencies higher than $1/T_2$. Increasing the system gain, with the system compensated by the lag network, effectively introduces a higher gain at the lower frequencies, thus decreasing the steady-state velocity error E_v. However, the phase margin θ_m must be kept the same (at very nearly the same phase margin frequency ω_m), so that the transient response is not altered. The phase lag introduced by the network must only be effective below the phase margin frequency. This means that as the greater phase change due to the compensator is between frequencies $1/T_2$ and A_2/T_2, the frequency $1/T_2$ must be much lower than the phase margin frequency ω_m.

The same criteria, as were used with Nyquist diagrams, can be used for obtaining the value of T_2, i.e. a small phase change at the system resonant frequency ω_r is specified. As previously, with $A_2 = 0{\cdot}2$ and $T_2 = 20$, the compensated system open-loop transfer function is

$$A\Phi_c(j\omega)\Phi_a(j\omega) = \frac{A0{\cdot}78}{j\omega(j\omega/2+1)(j\omega/3+1)}\,\frac{(j\omega 20+1)}{(j\omega 100+1)}$$

Figure 10.30 shows the Bode diagrams of the compensated system. We shall now compare quantities taken from the Bode diagrams of the compensated system with those from the uncompensated system. We shall also evaluate and compare steady-state errors. For direct comparison this is presented in a tabular form.

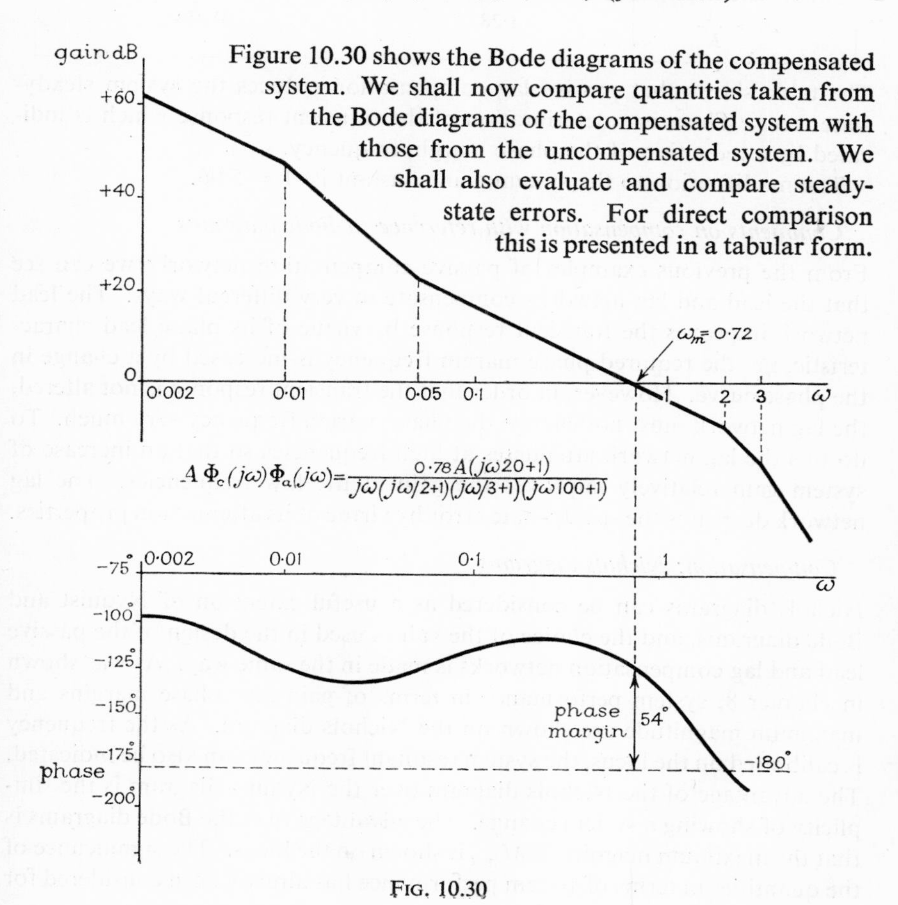

FIG. 10.30

Quantity	Uncompensated system $\Phi_a(j\omega) = \dfrac{0{\cdot}78}{j\omega(j\omega/2+1)(j\omega/3+1)}$	Compensated system passive lag compensation $A\Phi_c(j\omega)\Phi_a(j\omega) = \dfrac{A0{\cdot}78(j\omega 20+1)}{j\omega(j\omega/2+1)(j\omega/3+1)(j\omega 100+1)}$
Phase margin θ_m	54° (effective $\zeta = 0{\cdot}53$)	54° (effective $\zeta = 0{\cdot}53$)
Phase margin frequency ω_m	0·78 radian per second	0·72 radian per second
Steady-state error coefficient $K' = K_v$	0·78	$0{\cdot}78A = 3{\cdot}94$
Steady-state error $E_v = \dfrac{1}{K_v}$	1·28	0·254

The table shows that passive lag compensation reduces the system steady-state error. There is a deterioration of the transient response which is indicated by a reduction of the phase margin frequency.

The modification to the system gain constant is $A = 5{\cdot}06$.

Comments on compensation with reference to Bode diagrams

From the previous examples of passive compensation networks we can see that the lead and lag networks compensate in very different ways. The lead network improves the transient response by virtue of its phase lead characteristic, i.e. the required phase margin frequency is increased by a change in the phase curve. However, in order that the transient response is not altered, the lag network must not change the phase margin frequency very much. To do this the lag network attenuates at high frequencies so that an increase of system gain relatively improves the gain at the low frequencies. The lag network decreases the steady-state error by virtue of its attenuation properties.

Compensation: Nichols diagrams

Nichols diagrams can be considered as a useful extension of Nyquist and Bode diagrams, and the choice of the values used in the design of the passive lead and lag compensation networks is made in the same way. As was shown in chapter 8, system performance in terms of gain and phase margins and maximum magnitude are shown on the Nichols diagram. As the frequency is calibrated on the locus, the system resonant frequency can also be indicated. The advantage of the Nichols diagram over the Nyquist diagram is the simplicity of showing a system change. The advantage over the Bode diagrams is that the maximum magnitude M_{max} is shown on the locus. The significance of the quantities in terms of system performance has already been considered for the Nyquist and Bode diagrams, so the three examples are considered briefly.

Uncompensated system: Nichols diagram

The example of the uncompensated system has an open-loop transfer function

$$\Phi_a(j\omega) = \frac{0{\cdot}89}{j\omega(j\omega/2+1)(j\omega/3+1)}$$

The quantities indicating performance are: maximum magnitude $M_{max} = 1{\cdot}113$ (effective damping ratio $\zeta = 0{\cdot}53$), system resonant frequency $\omega_r = 0{\cdot}76$ radian per second and the phase margin $\theta_m = 55°$.

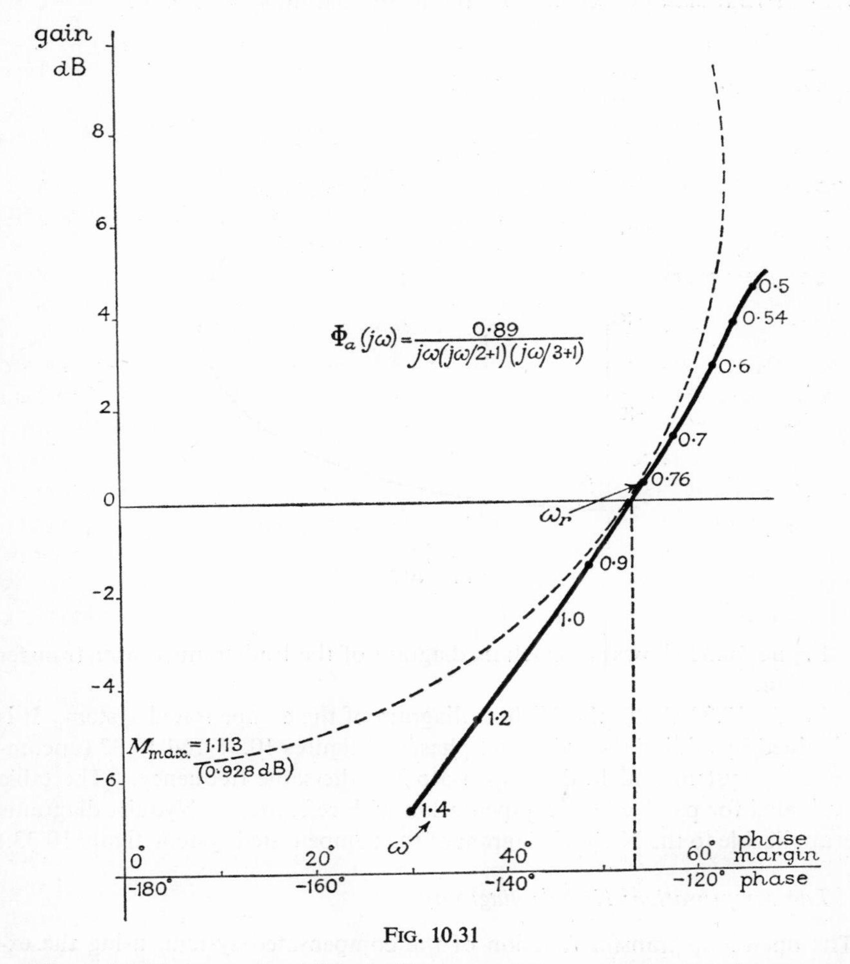

FIG. 10.31

Figure 10.31 shows the Nichols diagram of the uncompensated system.

Lead compensation: Nichols diagram

The open-loop transfer function of the compensated system, using the example of passive lead compensation, is

$$A\Phi_a(j\omega)\,\Phi_c(j\omega) = \frac{2{\cdot}13}{j\omega(j\omega/3+1)(j\omega/20+1)}$$

The quantities indicating performance are maximum magnitude $M_{max} = 1{\cdot}113$ (comparable with the effective damping ratio), system resonant frequency $\omega_r = 1{\cdot}72$ radian per second and the phase margin $\theta_m = 55{\cdot}25°$.

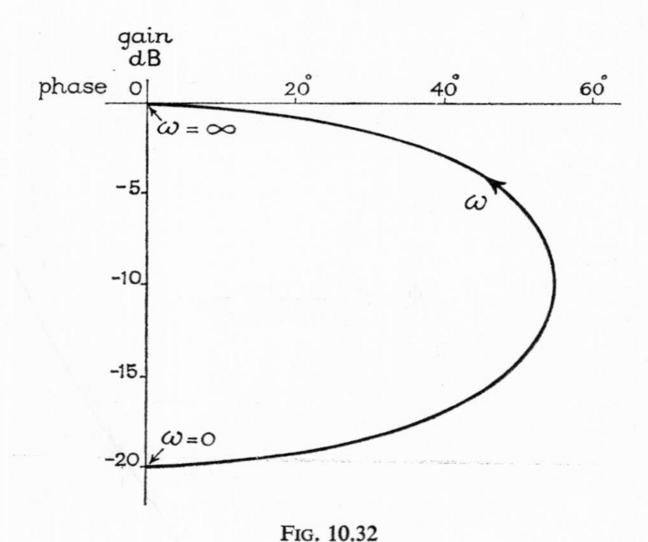

FIG. 10.32

Figure 10.32 shows the Nichols diagram of the lead compensator transfer function.

Figure 10.33 shows the Nichols diagram of the compensated system. It is obtained by adding the gains and phases of figures 10.31 and 10.32 (uncompensated system and lead compensator) at the same frequency. (The table evaluated for passive lead compensation with reference to Nyquist diagrams is applicable to the Nichols diagram of the compensated system, figure 10.33.)

Lag compensation: Nichols diagrams

The open-loop transfer function of the compensated system, using the example of passive lag compensation, is

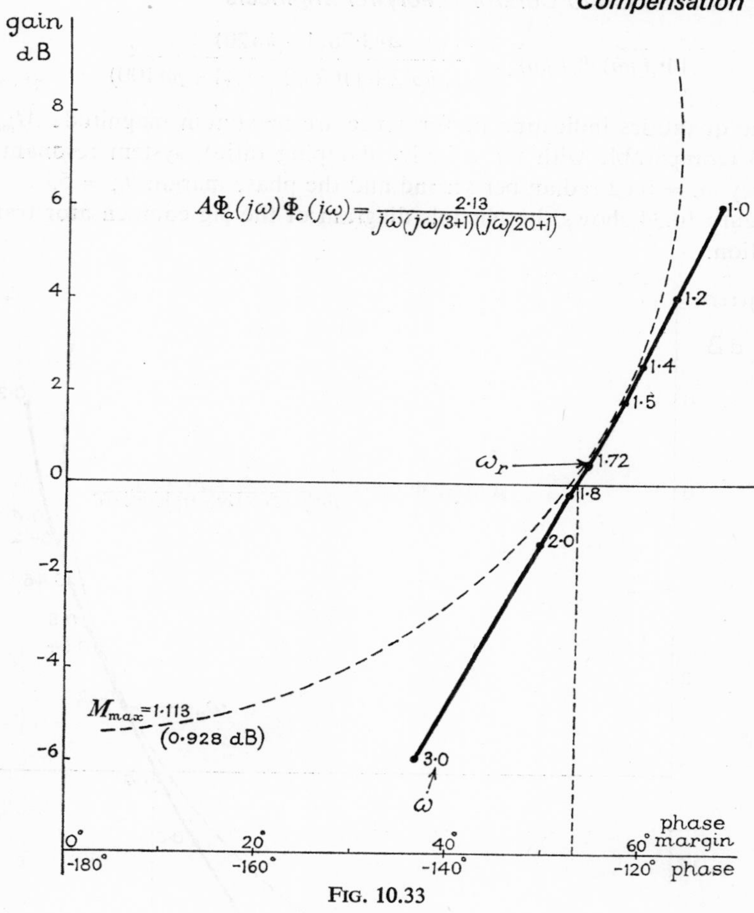

FIG. 10.33

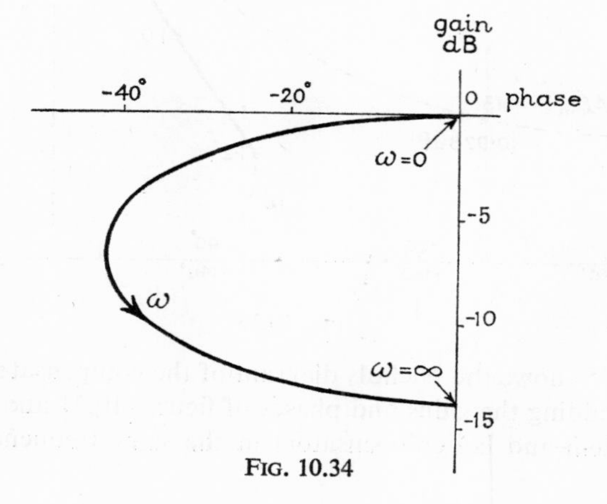

FIG. 10.34

$$A\Phi_a(j\omega)\,\Phi_c(j\omega) = \frac{3{\cdot}76(1+j\omega 20)}{j\omega(j\omega/2+1)(j\omega/3+1)(1+j\omega 100)}$$

The quantities indicating performance are maximum magnitude $M_{max} = 1{\cdot}113$ (comparable with the effective damping ratio), system resonant frequency $\omega_r = 0{\cdot}62$ radian per second and the phase margin $\theta_m = 53°$.

Figure 10.34 shows the Nichols diagram of the lag compensator transfer function.

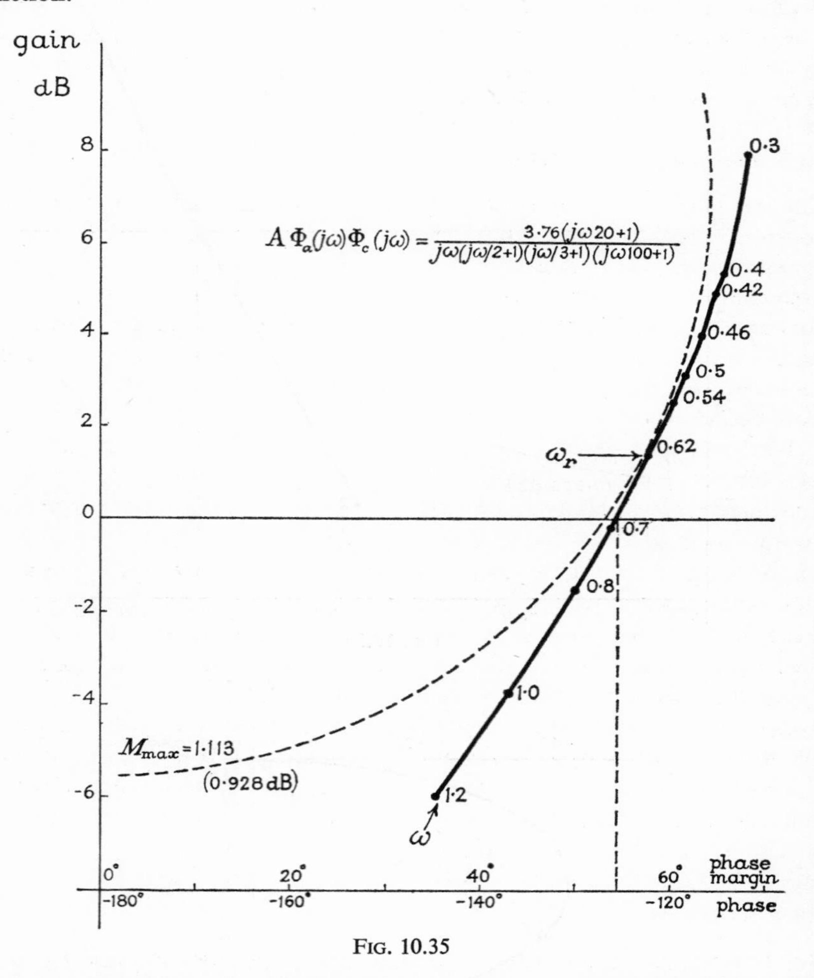

FIG. 10.35

Figure 10.35 shows the Nichols diagram of the compensated system. It is obtained by adding the gains and phases of figures 10.31 and 10.34 (uncompensated system and lag compensator) at the same frequency. (The table evaluated for passive lag compensation with reference to Nyquist diagrams is applicable to the Nichols diagram of the compensated system, figure 10.35.)

Comments

The Nichols diagram is particularly useful when a gain adjustment is required for a particular maximum magnitude M_{max}. A transparency of the M contours, gain calibration and phase calibration can be placed over a Nichols diagram and moved until the required M_{max} is indicated on the transparency. The difference in position of the transparency and the gain calibration on the Nichols diagram gives the gain adjustment (the transparency must be the same scale as the Nichols diagram calibration).

10.5 Feed-back compensation

The previous sections have dealt with the improvement of simple closed-loop control systems by the introduction of compensating elements into the forward path of the system. The elements compensate by modifying the system error. They can be introduced into the feed-back path instead of the forward path, and in this case modify the error by changing the feed-back output signal. Sometimes feed-forward compensation may not be practical and feed-back compensation can be employed. Similar elements are used in both cases.

Feed-back compensation has certain advantages. One is that the output of a control system is generally a high, power source compared with the input- and therefore the compensator does not need an amplifier (the feed-forward compensator necessitates an increased system gain). An example of the improvement of the same system by feed-forward and feed-back compensation is the introduction of damping in the remote position control system. Feed-forward improvement is viscous friction damping, and the equivalent feed-back improvement is velocity feed-back by means of a tachogenerator.

Another advantage of feed-back compensation is that less noise in the system results than with feed-forward compensation (the increased gain of the forward path with feed-forward compensation increases the noise level).

From the design point of view feed-back compensators are generally more difficult to apply than feed-forward compensators. More detailed discussion of compensation is outside the scope of this book.

10.6 Examples

1. Describe concisely, using a block diagram, the essential elements in an error-actuated automatic control system for the position control of a rotatable mass, and discuss briefly methods that may be used to make the system stable.

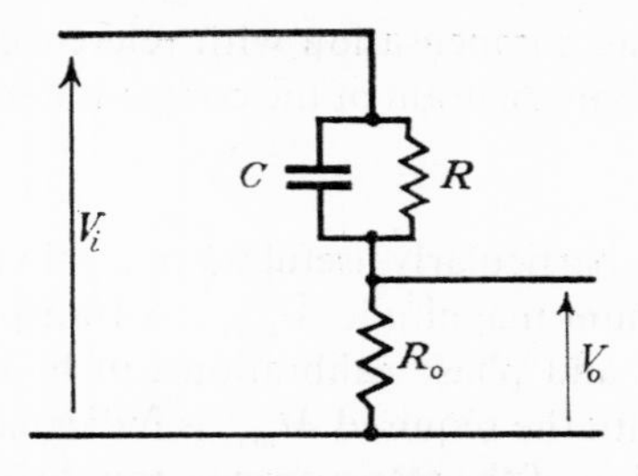

The circuit shown is used in an amplifier of a control system. Derive an expression for the transfer function of the circuit. If $V_i = 10 \sin 10t$ volts, $R = 50\,\text{k}\Omega$, $R_0 = 5\,\text{k}\Omega$ and $C = 1\,\mu\text{F}$, calculate the output voltage in magnitude and phase relative to V_i.

(Part 3, I.E.E., 1965)

$$\left[0{\cdot}091\,\frac{(1+s0{\cdot}05)}{(1+s0{\cdot}0046)};\quad 0{\cdot}91 \sin(10t-0{\cdot}054)\right]$$

2. Discuss what is meant by the term 'Design by Analysis' as applied to control systems.

3. Describe the principle and explain the behaviour of a friction-damped position control servo-mechanism with a controller torque (*a*) proportional to error and (*b*) proportional to error plus the derivative of error. Illustrate your desciiption in each case by a block schematic and set up the differential equation representing the performance.

(Part 3, I.E.E., 1956)

4. Explain what is meant by the steady-state error coefficient as applied to control systems. Find the steady-state error coefficient and the steady-state error of the control systems whose open-loop transfer functions are:

(*i*) $$\bar{\Phi}_a = \frac{10}{s(s+2)}$$

(*ii*) $$\bar{\Phi}_a = \frac{3}{s(s+4)(s+5)}$$

(*iii*) $$\Phi_a(j\omega) = \frac{0{\cdot}1}{j\omega(j\omega 0{\cdot}2+1)}$$

(*iv*) $$\bar{\Phi}_a = \frac{0{\cdot}1(s+50)}{s^2(s+5)(s+1{\cdot}0)(s+2)}$$

$$(v)\quad \Phi_a(j\omega) = \frac{1+j\omega}{1+j\omega 0{\cdot}1}$$

(5; 0·15; 0·1; 0·5; 1;
0·2; 66·67; 10; 2; 1)

5. An error-actuated closed-loop control system is used to control, by means of a hand-wheel, the angular position of an electrically driven turntable. Draw a block diagram and develop, with a full explanation, the differential equation for the angular position of the turn-table following a movement of the hand-wheel.

Indicate the errors likely to arise with such a system and explain how these could be reduced by adding derivative and/or integral control to the arrangement.

(Part 3, I.E.E., 1965)

6. The closed-loop response of a type 1 unity feed-back control system is second order. If the maximum magnitude is $M_{max} = 1{\cdot}157$ and the system resonant frequency is $\omega_r = 7$ radians per second, find the system gain constant K and the undamped natural frequency.

(0·5; 9·9)

7. A type 1 control system has an open-loop transfer function $\Phi_a(j\omega)$,

$$\Phi_a(j\omega) = \frac{K'}{j\omega(1+j\omega)(1+j\omega 0{\cdot}2)}$$

Find the maximum magnitude for a system gain constant $K = 4{\cdot}35$, and hence find the damping ratio of the dominant roots.

(1·25; 0·45)

8. Give an example of a type 0 control system and draw its block diagram. Hence obtain the system transfer function.

9. A single-loop unity feed-back control system has a second-order characteristic equation, a damping ratio $\zeta = 0{\cdot}4$, and an undamped natural frequency $\omega_n = 10$ radians per second. Find the phase margin frequency and the phase margin.

(8·544; 43° 10′)

10. Describe the function of a passive lag compensation network and a passive lead compensation network. Discuss both networks in terms of their pole-zero patterns and their frequency response diagrams.

11. Design compensation networks to improve the transient response and decrease the steady-state error of the single-loop unity feed-back control system whose open-loop transfer function is

$$\overline{\Phi}_a = \frac{3}{s(s+1)(s+2)}$$

CHAPTER 11

Introduction to Non-Linear Systems

*11.1 Introduction

When a control system is described by a linear differential equation (or transfer function) it is assumed to be linear. All systems, however, contain non-linearities. The simple remote position control system has been considered linear but it should be appreciated that under very small and very large input conditions the system is non-linear. (With a very small input the error signal may be too small to provide a torque to overcome static friction, and with a very large input the amplifier may saturate.) We can say that the majority of systems which are called linear are very slightly non-linear over their range of operation, and (or) become non-linear at the outside limits of their operation (i.e. for very small and very large inputs as already mentioned). However, the linear descriptions and methods of analysis can be applied satisfactorily. There are systems which have non-linearities that cannot be approximated to linearities, and there are systems where design demands the introduction of non-linear elements.

A non-linear system is one for which the principle of superposition† is no longer true.

In general there are no accurate methods for designing and analysing non-linear systems. If, however, information is known about particular non-linearities certain techniques may be applied. Non-linearities can be placed in two categories: 'useful' and 'unavoidable'.

Useful non-linearities are non-linear elements which are introduced into a system because their characteristics enable the system to work more efficiently.

Friction, dead-zone (i.e. a range of system input for which there is no

† The principle of superposition: If a system has outputs $C = C_1, C = C_2 \ldots, C = C_n$ for inputs $R = R_1, R = R_2, \ldots, R = R_n$, respectively, then $C = (C_1 + C_2 + \ldots + C_n)$ is an output for the input $R = (R_1 + R_2 + \ldots + R_n)$.

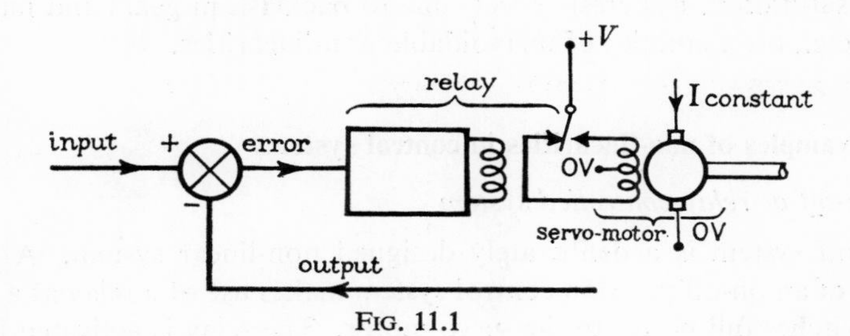

FIG. 11.1

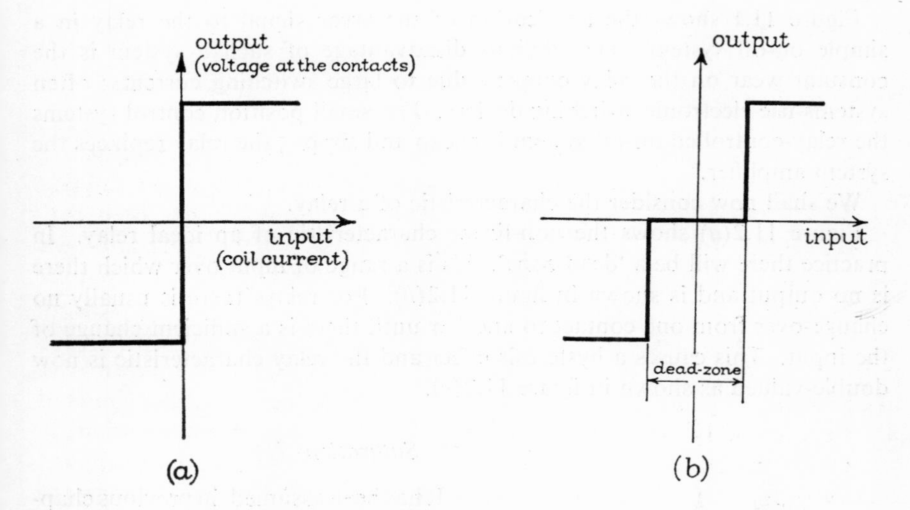

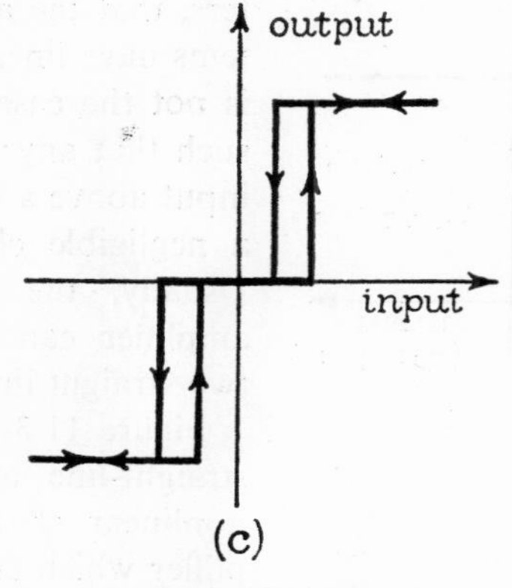

FIG. 11.2

output), saturation, hysteresis effects due to backlash in gears and magnetic circuits, etc., are examples of unavoidable non-linearities.

*11.2 Examples of non-linearities in control systems

The on-off or relay-controlled system

The on-off system is a deliberately designed non-linear system. A simple example of an on-off position control system makes use of a relay as a switch which switches full power to the servo-motor. The relay is activated by the error signal and the full correcting torque is applied only with respect to the sign of the error signal.

Figure 11.1 shows the application of the error signal to the relay in a simple on-off system. The obvious disadvantage of such a system is the constant wear on the relay contacts due to large switching currents; often systems use electronic switching devices. For small position control systems the relay-controlled on-off system is cheap and simple, the relay replaces the system amplifier.

We shall now consider the characteristic of a relay.

Figure 11.2(*a*) shows the non-linear characteristic of an ideal relay. In practice there will be a 'dead zone', this is a range of input over which there is no output and is shown in figure 11.2(*b*). For relays there is usually no change-over from one contact to another until there is a sufficient change of the input. This causes a hysteresis effect and the relay characteristic is now double-valued as shown in figure 11.2(*c*).

Saturation

It has been assumed, in previous chapters, that the amplifiers used in systems have linear characteristics; this is not the case. Amplifiers saturate such that any increase in a value of input above a certain level results in a negligible change in the output. Usually, the characteristic of an amplifier can be approximated by two straight lines.

Figure 11.3 shows the actual and straight-line approximation of the nonlinear characteristic of an amplifier which saturates for any value of input voltage greater than V_s.

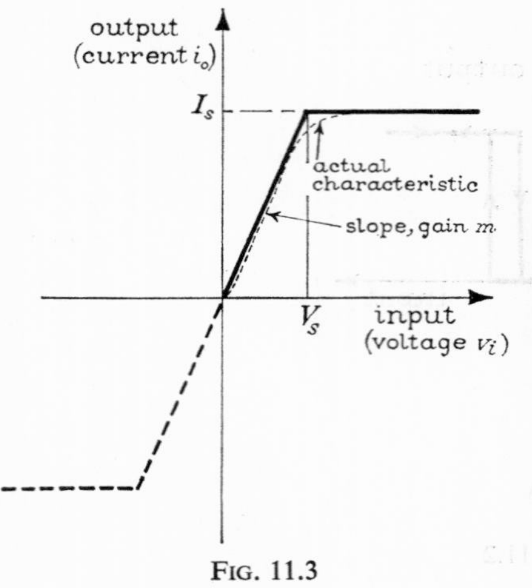

FIG. 11.3

Friction

Apart from the introduction of viscous friction in the remote position control system, the effects of friction have, so far, largely been ignored. In practice friction introduces non-linearities into a system. When a mechanical system is at rest an initial force is required to set it in motion; this force must overcome static friction (sometimes known as *stiction*). Once the system is in motion there is a constant rubbing friction irrespective of the speed of the system. This rubbing friction is less than the initial static friction. The static and rubbing friction are together known as Coulomb, or dry, friction. Coulomb friction is often considered a negligible effect.

11.3 The describing function

The describing function of a non-linear element enables a non-linear control system to be analysed using an approximate frequency response technique.

Let us consider a general single-loop unity feed-back non-linear control system.

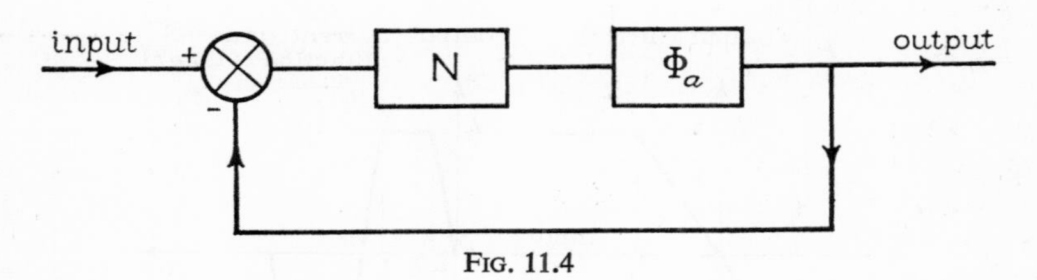

FIG. 11.4

Figure 11.4 shows the block diagram of the non-linear system, block N represents the non-linear part of the system and block Φ_a the linear part of the system. We shall consider a sinusoidal input to the block N on its own.

Figure 11.5 shows the non-linear block with an input $a = A \sin \omega t$; the non-linearity is assumed to be such that the output b is a function of the input amplitude only. The output is a periodic non-sinusoidal function and can be analysed into harmonic components by means of Fourier analysis. It is usually found, for control systems, that the predominant component of the output is the fundamental. In general, harmonics can be ignored because the linear block Φ_a attenuates them such that they have a much smaller amplitude than the fundamental. (The block Φ_a usually has a transfer func-

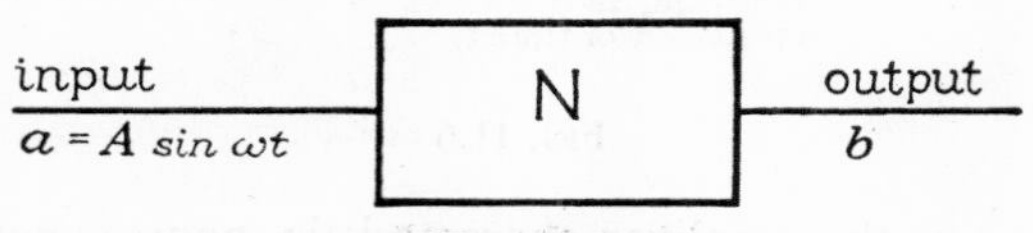

FIG. 11.5

tion whose denominator has a greater order of s than the numerator. This means that block Φ_a behaves like an integrating element to some extent ($1/s$ is the operation of integration). Integration decreases the amplitude of the harmonic components with respect to the fundamental, (for example a third-harmonic component $B_3 \cos 3\omega t$ becomes $\dfrac{B_3}{3\omega} \sin 3\omega t$, a decrease to a third of its amplitude with respect to the fundamental.)

The describing function $N(A)$, a function of the amplitude of the input only, is defined as

$$N(A) = \frac{\text{magnitude of the fundamental component of the output}}{\text{magnitude of the sinusoidal input}} \tag{11.1}$$

Output and input here refer to the output and input of the non-linear element N.

To illustrate the describing function we shall consider the example of the saturating amplifier.

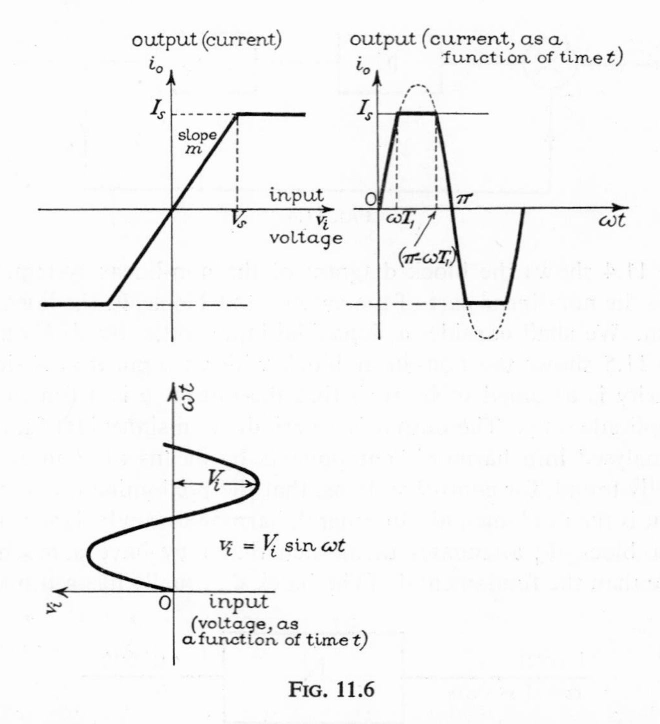

FIG. 11.6

Figure 11.6 shows the amplifier characteristic, approximated by straight lines. The amplifier can be considered to have a gain m until saturation. The input voltage is $v_i = V_i \sin \omega t$; the output current is i_o, where

$$i_o = I_o \sin \omega t, \quad \text{for} \quad 0 < \omega t < \omega T_1$$

$$\text{and} \qquad i_o = I_s \quad , \quad \text{for} \quad \omega T_1 < \omega t < \pi - \omega T_1$$

To obtain the fundamental component of the output we shall use Fourier analysis. By inspection the output contains only sinusoidal components and only odd harmonics. The nth harmonic, because of the symmetry of the output, can be expressed by

$$I_{o_n} = \frac{2}{\pi} \int_0^{\pi} i_o \sin n\omega t \, d(\omega t)$$

where $n = 1, 2, 3,$ etc. The fundamental component is found by evaluating the previous integral for $n = 1$.

$$I_{o_1} = \frac{2}{\pi} \left[I_o \left(\omega T_1 - \frac{\sin 2\omega T_1}{2} \right) + 2 I_s \cos \omega T_1 \right]$$

but $I_s/V_s = m$ and $I_o/V_i = m$, therefore the describing function, assuming saturation occurs, is

$$N(A) = \frac{I_{o_1}}{V_i} = \frac{2m}{\pi} \left[\omega T_1 - \frac{2 \cos \omega T_1 \sin \omega T_1}{2} + 2 \frac{V_s}{V_i} \cos \omega T_1 \right] \qquad (11.2)$$

when $\omega t = \omega T_1$, $V_s = V_i \sin \omega T_1$. Substituting for ωT_1 in equation 11.2 we have

$$N(A) = \frac{2m}{\pi} \left[\sin^{-1} \frac{V_s}{V_i} - \frac{V_s}{V_i} \cos \left(\sin^{-1} \frac{V_s}{V_i} \right) + 2 \frac{V_s}{V_i} \cos \left(\sin^{-1} \frac{V_s}{V_i} \right) \right]$$

$$N(A) = \frac{2m}{\pi} \left[\sin^{-1} \frac{V_s}{V_i} + \frac{V_s}{V_i} \cos \left(\sin^{-1} \frac{V_s}{V_i} \right) \right] \qquad (11.3)$$

The describing function for a saturating amplifier, in terms of the amplitude of the input, is equation 11.3.

The describing function $N(A)$ has a different value for every value of amplitude of the input; it is the 'gain' of the amplifier for a particular sinusoidal input. It can be considered as being somewhat similar to a transfer function (successive describing functions in a path of a control system can be added together to give an overall describing function).

We shall now briefly examine how, using the describing function of a non-linear control system, we can obtain some information about the system. The describing function is essentially a frequency response technique and we shall use a modified Nyquist locus to give us information about the non-linear system. Using the describing function $N(A)$ in the same way as a transfer function, the closed-loop transfer function $\Phi_a(j\omega)$ of the general single-loop unity feed-back non-linear control system is given by

$$\Phi(j\omega) = \frac{N(A)\Phi_a(j\omega)}{1+N(A)\Phi_a(j\omega)} \tag{11.4}$$

To investigate the stability of a system the Nyquist condition is used. For the system described by equation 11.4 the locus of $N(A)\Phi_a(j\omega)$ must not enclose the $(-1, j0)$ point for the system to be stable. We shall consider describing functions that are a function of amplitude, but independent of frequency. This means for a particular gain constant many Nyquist loci of $N(A)\Phi_a(j\omega)$ can be drawn because the describing function $N(A)$ has a different value for every value of input amplitude. These Nyquist loci give the condition of stability for the system for each value of the input amplitude. However, R. J. Kochenberger developed a better representation, which consists of drawing separate loci for the linear and non-linear 'blocks' of the control system. Consider the locus of $N(A)\Phi_a(j\omega)$: if it intersects the negative real axis at the $(-1, j0)$ point, by Nyquist's condition, the closed-loop system output is a constant-amplitude oscillation. Therefore

$$N(A)\Phi_a(j\omega) = -1$$

$$\Phi_a(j\omega) = \frac{-1}{N(A)} \tag{11.5}$$

Equation 11.5 indicates that two loci can be drawn: one of $\Phi_a(j\omega)$ as for a linear system, the other of $-1/[N(A)]$ for all possible values of input amplitude. Where these loci intersect-equation 11.5 is satisfied and the closed-loop system output is a constant-amplitude oscillation. Equation 11.5 suggests that the 'describing function locus' $-1/[N(A)]$ can be considered as the locus of the $(-1, j0)$ point ($-1/[N(A)]$ will be specified as the describing function locus even though it is actually the inverse of the describing function). The locus of $-1/[N(A)]$ is calibrated in amplitude A, where A represents the relevant function of amplitude of the input to the non-linear block N. If the $\Phi_a(j\omega)$ locus totally encloses the $-1/[N(A)]$ locus then a completely unstable system is indicated. However, if the $-1/[N(A)]$ locus is not enclosed, or intersected by the $\Phi_a(j\omega)$ locus, the system is completely stable.

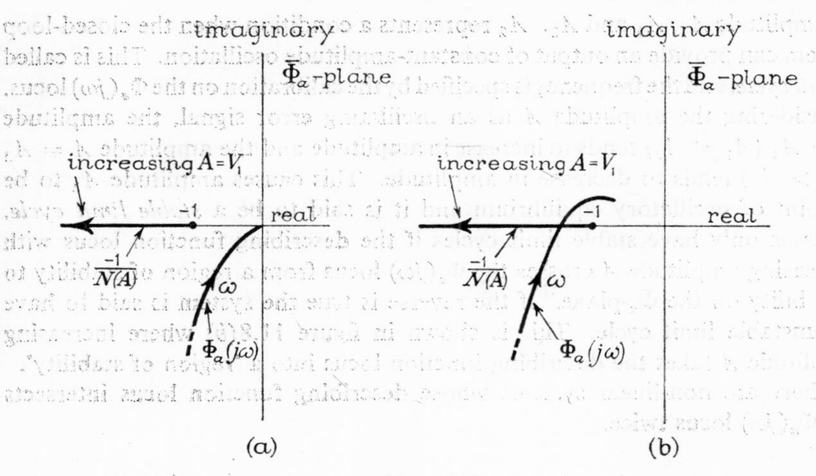

FIG. 11.7

Figure 11.7(*a*) and (*b*) show the loci for non-linear systems, the non-linearity being a saturating amplifier (figure 11.6). Figure 11.7(*a*) shows a completely stable non-linear system. Figure 11.7(*b*) shows a non-linear system which is unstable for a certain value of the input signal to the amplifier (the error signal).

We shall now examine more thoroughly non-linear systems whose describing function locus intersects the $\Phi_a(j\omega)$ locus.

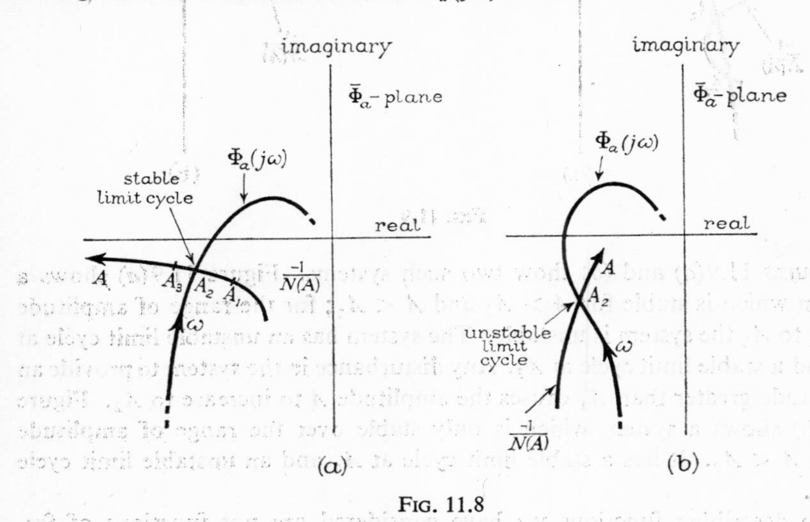

FIG. 11.8

Figure 11.8(*a*) shows a describing function locus calibrated for three values of amplitude A_1, A_2 and A_3. A_2 represents a condition when the closed-loop system can provide an output of constant-amplitude oscillation. This is called a *limit cycle*, and the frequency is specified by the calibration on the $\Phi_a(j\omega)$ locus. Considering the amplitude A as an oscillating error signal, the amplitude $A = A_1$ ($A_1 < A_2$) tends to increase in amplitude and the amplitude $A = A_3$ ($A_3 > A_2$) tends to decrease in amplitude. This causes amplitude A_2 to be a point of oscillatory equilibrium and it is said to be a *stable limit cycle.* Systems only have stable limit cycles if the describing function locus with increasing amplitude A crosses the $\Phi_a(j\omega)$ locus from a region of stability to instability on the $\overline{\Phi}_a$-plane. If the reverse is true the system is said to have an unstable limit cycle. This is shown in figure 11.8(*b*) where increasing amplitude A takes the describing function locus into a 'region of stability'.

There are non-linear systems whose describing function locus intersects the $\Phi_a(j\omega)$ locus twice.

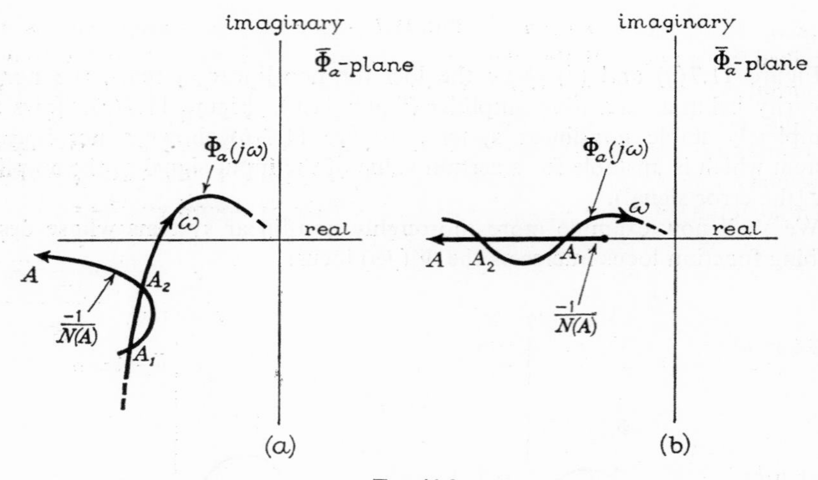

FIG. 11.9

Figures 11.9(*a*) and (*b*) show two such systems. Figure 11.9(*a*) shows a system which is stable for $A > A_2$ and $A < A_1$; for the range of amplitude A, A_1 to A_2 the system is unstable. The system has an unstable limit cycle at A_1 and a stable limit cycle at A_2. Any disturbance in the system to provide an amplitude greater than A_1 causes the amplitude A to increase to A_2. Figure 11.9(*b*) shows a system which is only stable over the range of amplitude $A_1 < A < A_2$. It has a stable limit cycle at A_1 and an unstable limit cycle at A_2.

The describing functions we have considered are not functions of frequency ω. If a system has a describing function which is dependent on amplitude and frequency, a series of describing function loci must be plotted for a range of frequency. This technique is complicated.

11.4 The phase-plane method

The behaviour of systems under given conditions can be described by various methods. One of these is a graph of the first derivative of the dependent variable (system output) of the system characteristic differential equation. This is only useful for second-order linear and non-linear systems. In order to define various singularities and to develop the phase-plane method we shall first consider linear second-order systems.

The characteristic equation of an undamped second-order linear remote position control system is

$$\ddot{\theta}_o + \omega_n^2 \theta_o = 0 \tag{11.6}$$

using the notation $d^2\theta_o/dt^2 = \ddot{\theta}_o$. The motion of this system can be described by a graph of the velocity of the output $d\theta_o/dt = \dot{\theta}_o$ to the base of output position. The system is assumed to be initially at rest, the output and input positions being identical, but displaced from zero. At time $t = 0$ seconds the input is brought to zero position and the graph of the output velocity θ_o is plotted. Equation 11.6 can be rewritten as a first-order differential equation eliminating time t. This is done by dividing equation 11.6 by

$$\frac{d\theta_o}{dt} = \dot{\theta}_o \tag{11.7}$$

Equation 11.6 may be expressed as

$$\frac{d\dot{\theta}_o}{dt} = -\omega_n^2 \theta_o$$

Dividing by equation 11.7, we obtain

$$\frac{d\dot{\theta}_o}{d\theta_o} = -\omega_n^2 \frac{\theta_o}{\dot{\theta}_o} \tag{11.8}$$

therefore

$$\dot{\theta}_o d\dot{\theta}_o = -\omega_n^2 \theta_o d\theta_o$$

Integrating both sides, we obtain

$$\dot{\theta}_o^2 = -\omega_n^2 \theta_o^2 + \text{constant} \tag{11.9}$$

Equation 11.9 represents the system in terms of output velocity and position and the required graph can be plotted. It represents a series of ellipses whose sizes are dependent on the initial displacement (giving the constant of integration).

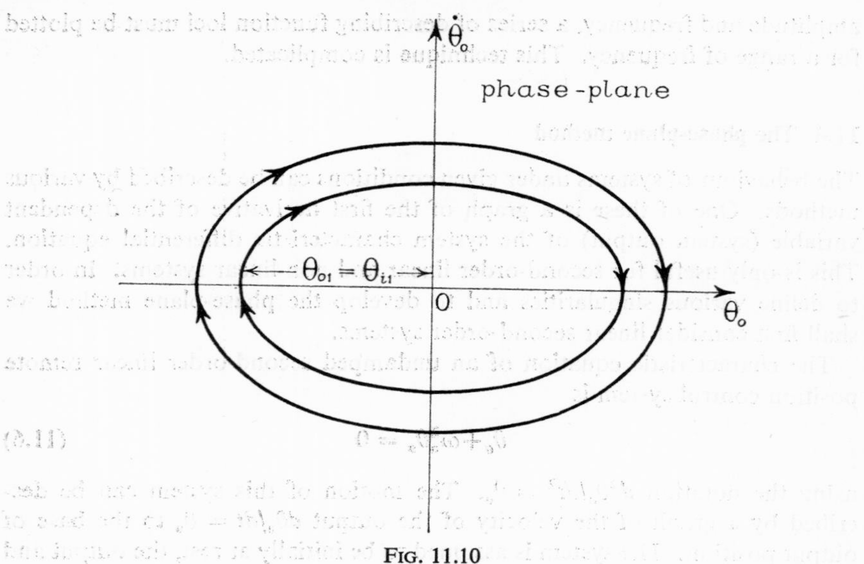

FIG. 11.10

Figure 11.10 shows two ellipses. Initially the system is at rest such that the output and input positions are the same, $\theta_{o_1} = \theta_{i_1}$. The input θ_{i_1} is then brought to zero at time $t = 0$ and the graph of output velocity $\dot{\theta}_o$ to the base of output position is as shown. The ellipse is to be expected as we know that the undamped simple remote position control system has a constant-amplitude oscillating output.

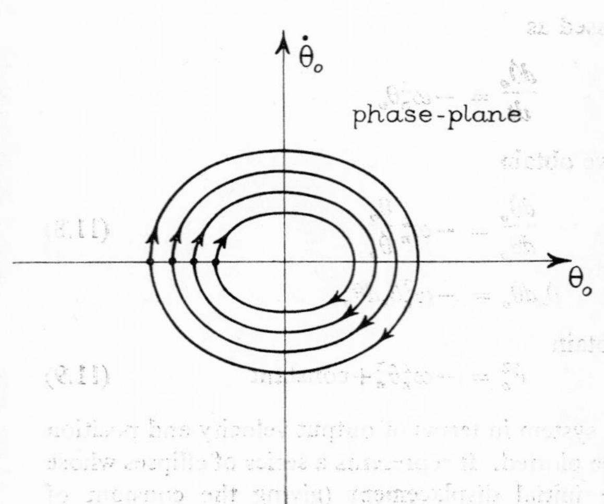

FIG. 11.11

The plane for plotting the first derivative of the dependent variable of the system to the base of the dependent variable is called the *phase-plane*. For each initial condition there is one path on the phase-plane; each path is called a *trajectory*. A family of trajectories on the phase-plane is called a *phase portrait*.

Figure 11.11 shows the phase portrait of equation 11.6. A periodic output always shows a closed curve on the phase-plane.

We shall now examine the phase portrait of an underdamped linear second-order system. For the remote position control system the characteristic equation is

$$\ddot{\theta}_o + 2\zeta\omega_n\dot{\theta}_o + \omega_n^2\theta_o = 0$$

however, for simplicity we shall consider the general equation

$$\ddot{x} + a\dot{x} + bx = 0 \tag{11.10}$$

where a and b are constants and x is the dependent variable. As previously

$$\frac{dx}{dt} = \dot{x} \tag{11.11}$$

and

$$\frac{d\dot{x}}{dt} = -a\dot{x} - bx \tag{11.12}$$

Dividing equation 11.12 by equation 11.11, we obtain

$$\frac{d\dot{x}}{dx} = -a - \frac{bx}{\dot{x}} \tag{11.13}$$

To construct a trajectory on the phase-plane we shall use the *method of isoclines*. This consists of constructing lines on the phase-plane, where $d\dot{x}/dx$ (that is the slope of the trajectories) is constant. These lines are called *isoclines*. If we substitute $d\dot{x}/dx$ = a constant k into equation 11.13 we obtain the equation of the isocline k.

$$k = -a - \frac{bx}{\dot{x}}$$

therefore

$$\frac{\dot{x}}{x} = \frac{-b}{k+a} \tag{11.14}$$

Using equation 11.14, lines of slope $-b/(k+a)$ can be constructed on the phase-plane and these are the isoclines specified by k. For instance when $k = 0$ the isocline is the line of slope $-b/a$ going through the origin; this indicates that the slope of the trajectories passing through this isocline all have zero slope. Similarly, if $k = \infty$ the isocline has a zero slope (it is the x axis).

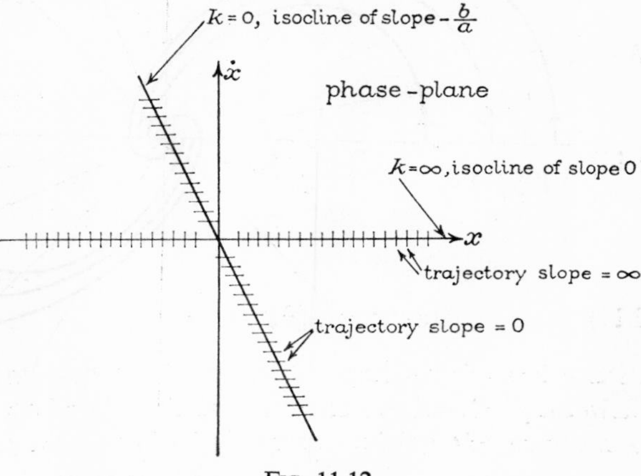

FIG. 11.12

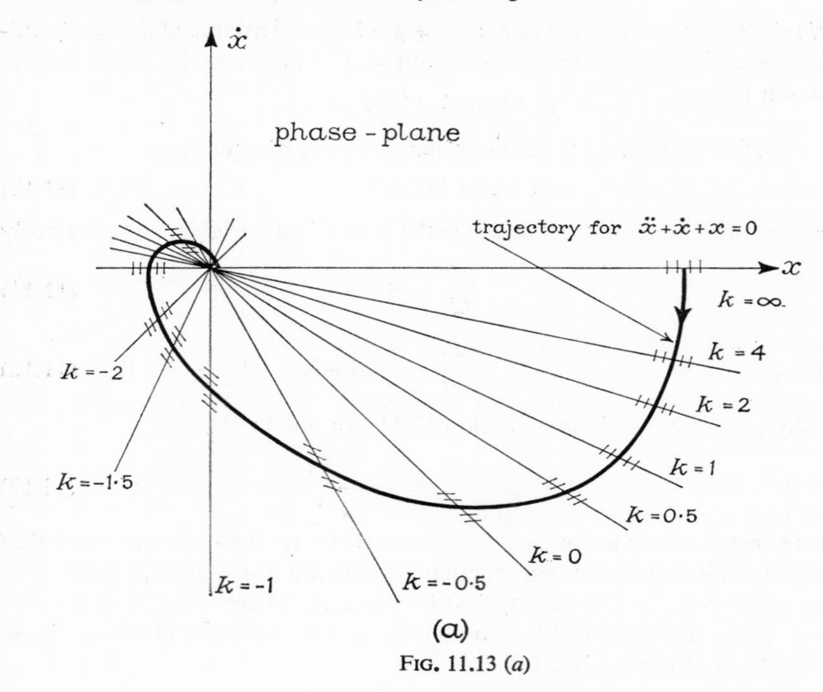

FIG. 11.13 (a)

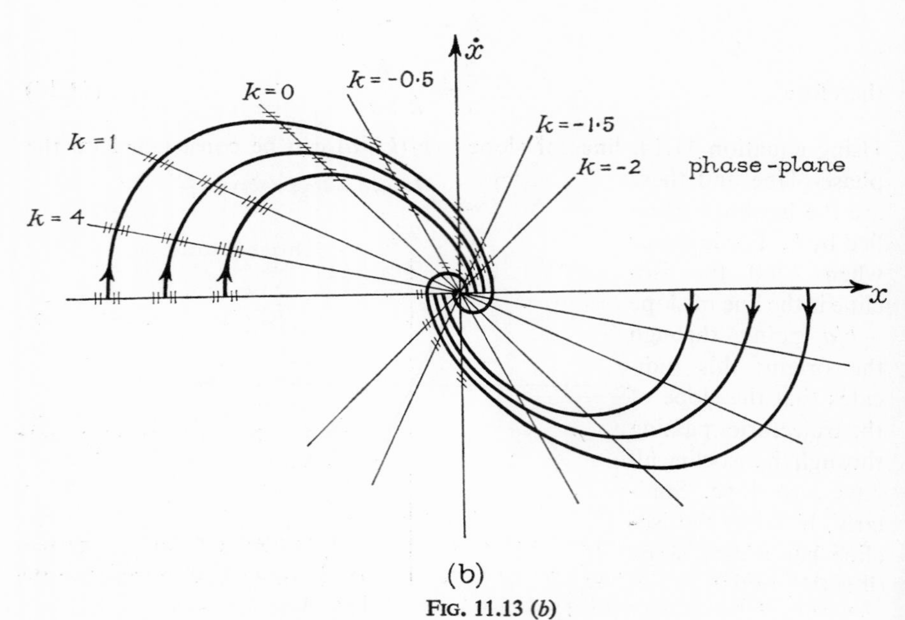

FIG. 11.13 (b)

Figure 11.12 shows the isoclines for $k=0$ and $k=\infty$ and sections of the infinite number of trajectories cutting these isoclines.

A trajectory can be sketched using a large number of isoclines.

Figure 11.13(*a*) shows a trajectory of equation 11.10 for the condition $0 < a^2/4b < 1$. This corresponds to the underdamped linear second-order remote position control system, $0 < \zeta < 1$. The trajectory shows that the system is damped oscillatory eventually coming to rest. Figure 11.13(*b*) shows the system phase portrait.

To interpret the phase portrait we define various *singular points* on the phase plane. These are points through which the slope of the trajectories is indeterminant, representing an equilibrium in a dynamic system. The singular points we shall consider are at the origin of the phase-plane and we shall discuss them in terms of the roots of the characteristic equation 11.10.

(*i*) *Imaginary roots*

For only imaginary roots ($a = 0$) we have an equation of the form of equation 11.6, the singular point, for dynamic equilibrium $x = \dot{x} = 0$, is at the centre of the phase-plane. The origin of the phase-plane is surrounded by the phase portrait of ellipses and is called a *centre*.

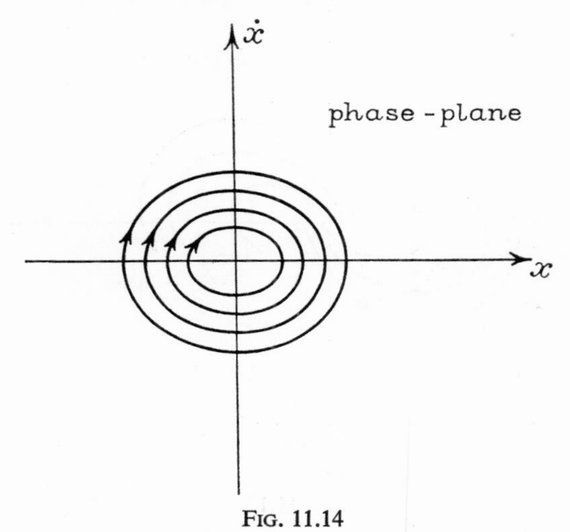

FIG. 11.14

Figure 11.14 shows a centre; the system output is of constant amplitude oscillation.

(*ii*) *Complex conjugate roots*

For complex conjugate roots the singular point is called a *focus*. If the real parts of the roots are negative the focus is called a *stable focus*, if positive the focus is called an *unstable focus*.

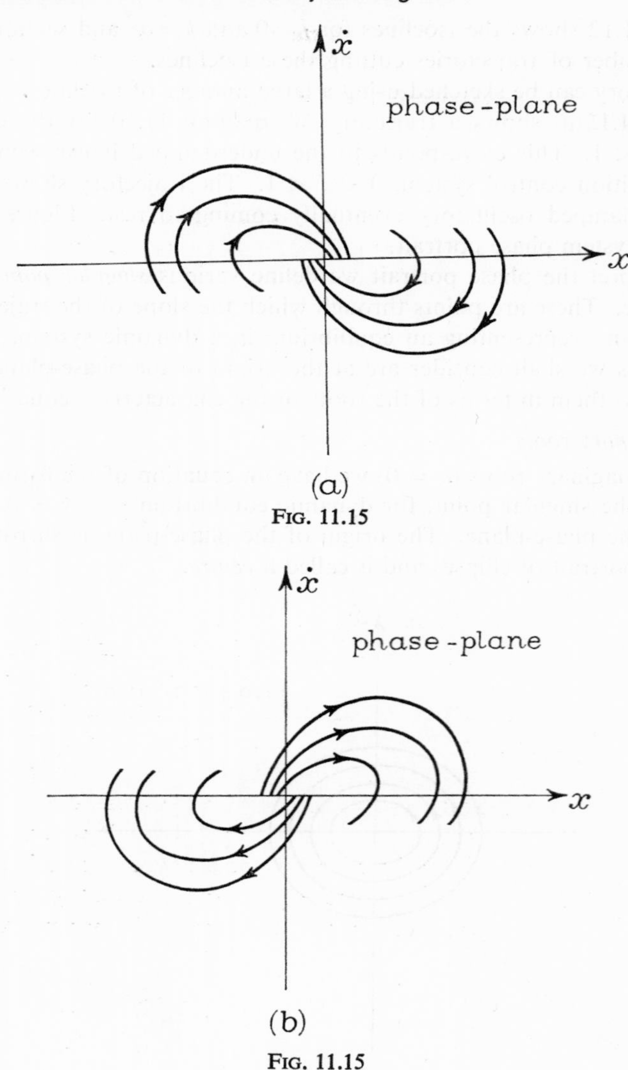

FIG. 11.15

FIG. 11.15

Figure 11.15(*a*) shows a stable focus and figure 11.15(*b*) shows an unstable focus. A stable focus represents a system output which is damped oscillatory, an unstable focus represents a system output which is oscillatory and exponentially increasing.

(*iii*) *Real roots*

For real roots of the same sign the singular point is called a *node*. If the real roots are negative the node is called a *stable node*, if the real roots are positive the node is called an *unstable node*. If the real roots are of opposite sign then the singular point is called a *saddle point*.

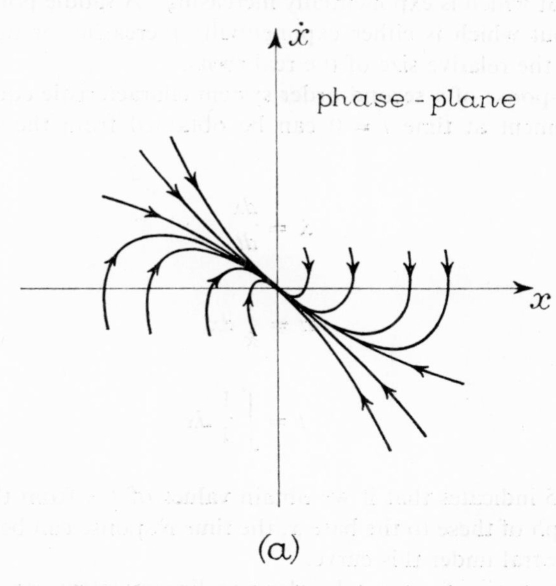

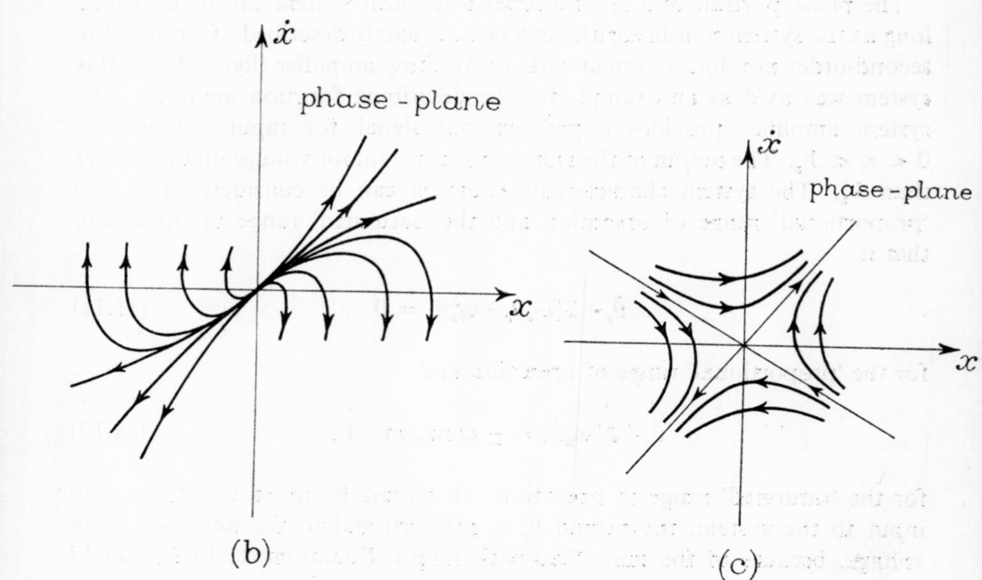

FIG. 11.16

Figures 11.16(a), (b) and (c) show a stable node, an unstable node and a saddle point respectively. A stable node represents a system output which is exponentially decreasing until it comes to rest. An unstable node represents a system output which is exponentially increasing. A saddle point represents a system output which is either exponentially increasing or decreasing depending upon the relative size of the real roots.

The time response of a second-order system characteristic equation for an input displacement at time $t = 0$ can be obtained from the trajectory as follows:

$$\dot{x} = \frac{dx}{dt}$$

therefore

$$dt = \frac{1}{\dot{x}}\,dx$$

and

$$t = \int \frac{1}{\dot{x}}\,dx \tag{11.15}$$

Equation 11.15 indicates that if we obtain values of $1/\dot{x}$ from the trajectory and plot a graph of these to the base x, the time response can be obtained by finding the integral under this curve.

The phase portrait of a second-order non-linear system can be drawn as long as the system non-linearities can be adequately described. Consider the second-order non-linear system with a saturating amplifier (figure 11.6) (this system was used as an example for the describing function method). The system amplifier provides a proportional signal for input voltages v_i, $0 < v_i < V_s$. The output of the amplifier is constant for voltage inputs greater than V_s. The system characteristic equation can be considered over the 'proportional' range of operation and the 'saturated' range of operation, that is

$$\ddot{\theta}_o + 2\zeta\omega_n\dot{\theta}_o + \omega_n^2\theta_o = 0 \tag{11.16}$$

for the 'proportional' range of operation and

$$\ddot{\theta}_o + 2\zeta\omega_n\dot{\theta}_o = \pm\text{constant} \times V_s \tag{11.17}$$

for the 'saturated' range of operation. (It should be noted that for a zero input to the system, the output θ_o is proportional to the amplifier input voltage, because of the unity feed-back loop.) Equations 11.16 and 11.17 indicate that the phase-plane should be split up into three sections according to whether $\pm\theta_o$ is greater than or less than $\pm$constant $\times V_s$. The three sections are shown on figure 11.17. The isoclines for equation 11.16 are lines radial at the origin, they are given by

$$\frac{\dot{\theta}_o}{\theta_o} = \frac{-\omega_n^2}{k_1+2\zeta\omega_n} \tag{11.18}$$

The isoclines for equation 11.17 are given by the straight lines

$$\dot{\theta}_o = \frac{\pm\text{constant} \times V_s}{k_2+2\zeta\omega_n} \tag{11.19}$$

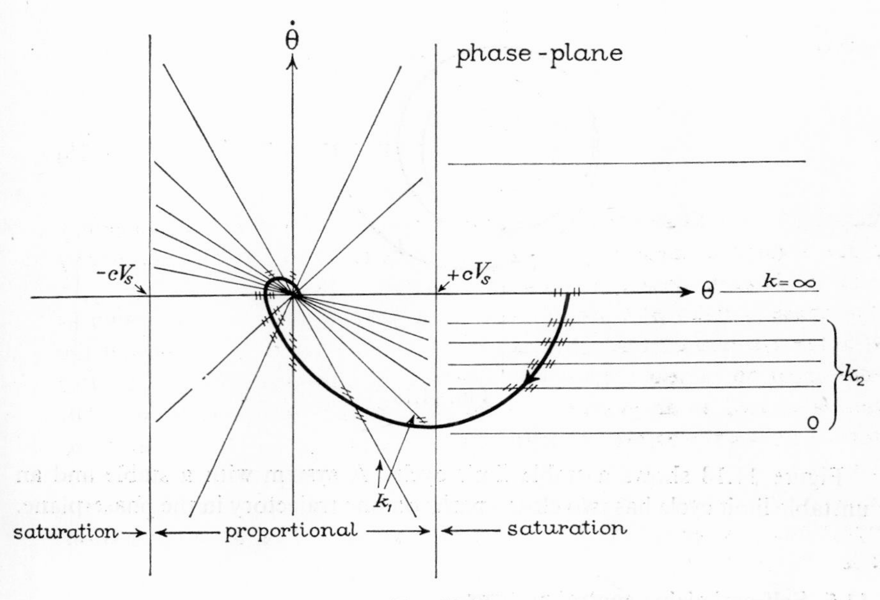

FIG. 11.17

Figure 11.17 shows a sketch of a typical trajectory of a second-order non-linear system with a saturating amplifier. The trajectory is drawn in the same way as for the linear systems, each region of the phase-plane being considered separately. Equations 11.18 and 11.19 give the isoclines for the regions of 'proportional' and 'saturated' operation respectively. A sharp transition from linear to saturation characteristic is assumed here; in practice the amplifier characteristic is curved and there is no abrupt boundary on the phase-plane. However, the trajectory shown is a useful approximation.

Limit cycles

In the previous section on describing functions, non-linear systems which appear to be unstable with an increasing oscillatory output, but which eventually settle down with a constant-amplitude oscillation called a stable limit cycle, were discussed. We have seen in this section that a closed trajectory on the phase-plane indicates a constant-amplitude oscillation, therefore we expect stable and unstable limit cycles to appear as closed trajectories on the phase-plane.

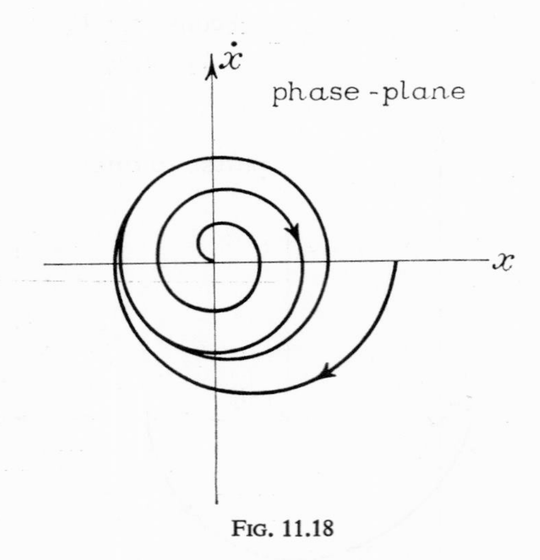

FIG. 11.18

Figure 11.18 shows a stable limit cycle. A system with a stable and an unstable limit cycle has two closed paths on one trajectory in the phase-plane.

11.5 Self-optimizing control systems

The optimum performance of a control system is defined in terms of the maximum error a system can tolerate to provide an accurate output. A *self-optimizing* control system (sometimes called an *adaptive* or *self-adaptive* control system) is a system which alters itself according to the change in environment and the changes within the system itself. It is a control system in which the parameters of the elements can effectively be changed to provide the optimum performance during the operation of the system.

A self-optimizing control system can make use of an ideal or 'model' system.

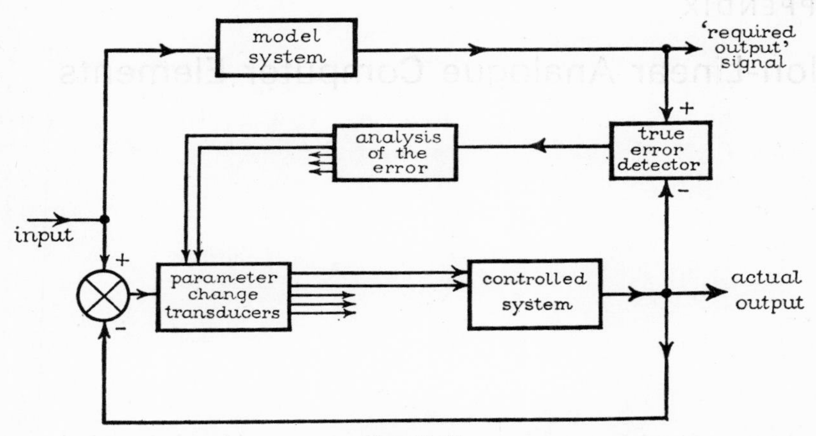

FIG. 11.19

Figure 11.19 shows a block diagram of a self-optimizing control system which has two error detectors, one of which is a 'true' error detector which compares the actual output with the output of a 'model' system. The 'model' system could be an analogue computer whose transfer function gives a defined optimum performance over the range of output required.

APPENDIX

Non-Linear Analogue Computer Elements

In chapters 5 and 6 on analogue computing we considered the solution and simulation of linear systems. There are various non-linearities which can be simulated using d.c. amplifiers and diodes; the following sections describe some of them.

a.1 Saturation

An analogue element which simulates saturation is called a *limiter*.

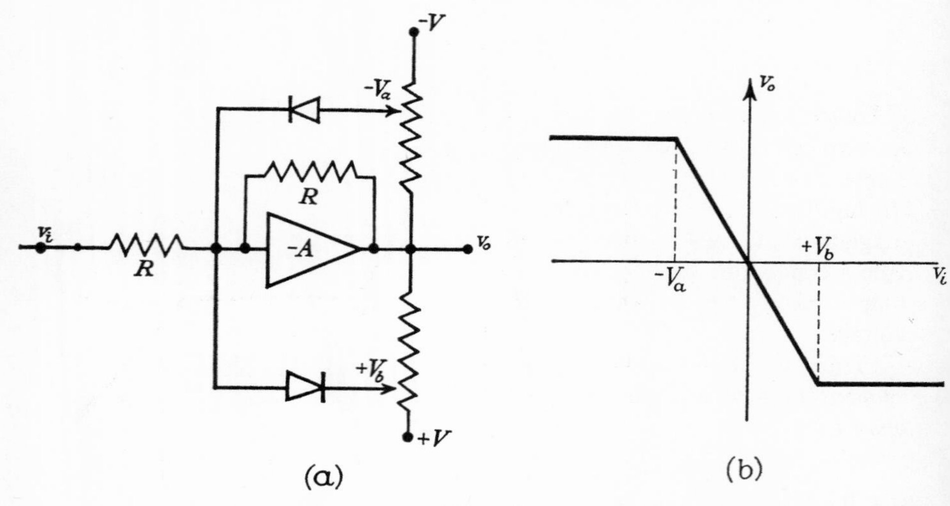

FIG. a.1

Figure a.1 (*a*) shows a limiter which gives a nearly constant output voltage v_o for any value of input voltage v_i which makes one of the diodes conduct, that is $+v_i > +V_b$ and $-v_i > -V_a$. Figure a.1 (*b*) shows the limiter characteristic which simulates saturation.

a.2 Dead-zone

A system is said to have a *dead-zone* when over a range of its input there is no output.

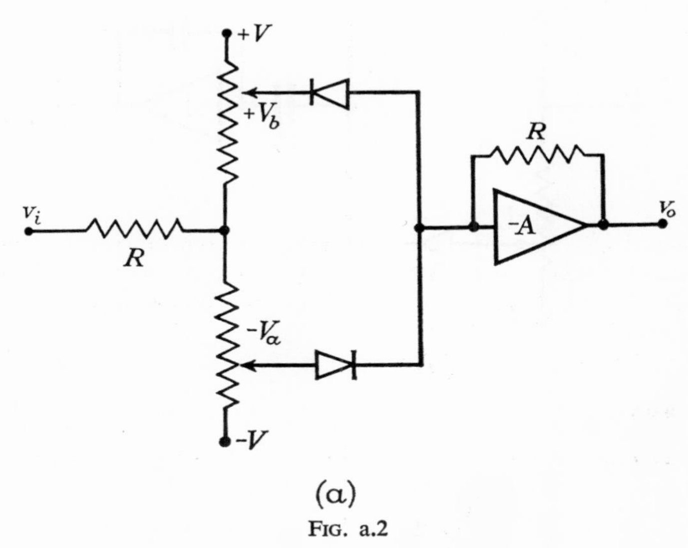

FIG. a.2

Figure a.2(*a*) shows an analogue element which simulates dead-zone. There is only an output voltage if the input voltage has a large enough magnitude to enable a diode to provide a conducting path. There is no output voltage for the range of input voltage.

Figure a.2(*b*) shows the analogue element characteristic which simulates dead-zone.

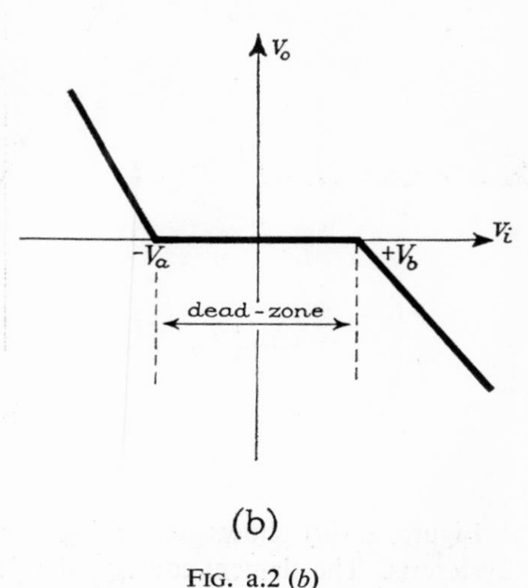

FIG. a.2 (*b*)

a.3 Backlash and hysteresis

Backlash or hysteresis occur in a system when the output of the system is double-valued for a cyclic input; that is, there can be two values of output for one value of input depending upon whether the input is increasing or decreasing.

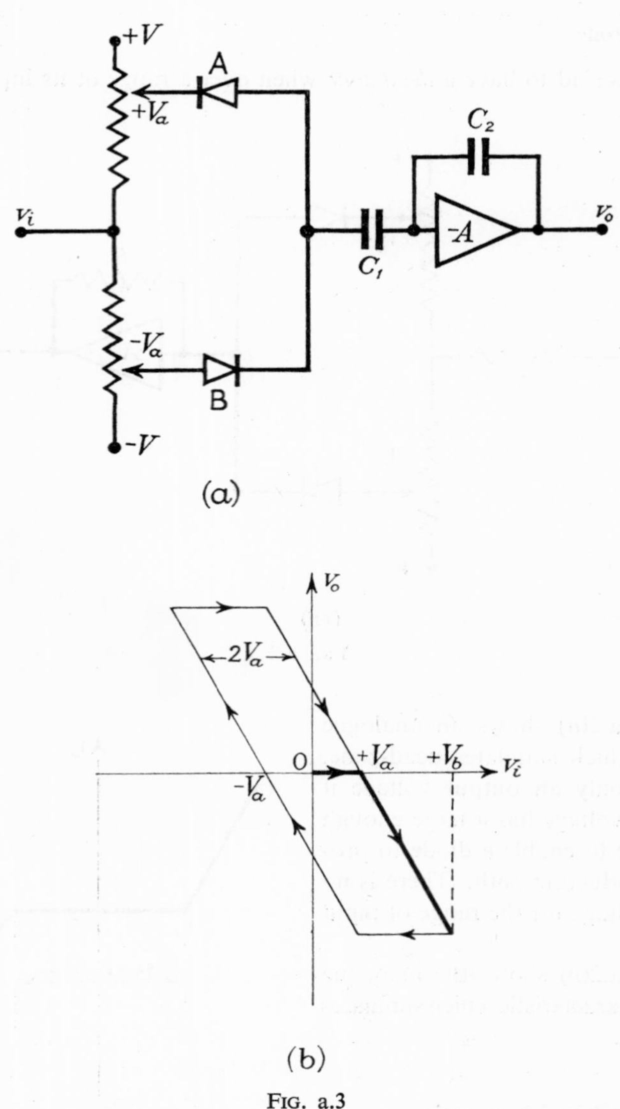

FIG. a.3

Figure a.3(*a*) shows an analogue element which simulates backlash or hysteresis. The element consists of a dead-zone circuit followed by an operational amplifier of transfer function $-C_2/C_1$. When the input voltage is increased from zero to $+V_b$ the characteristic, figure a.3(*b*), is the same as the dead-zone. If the input voltage is now decreased the diode (A) no longer conducts, the capacitors are unable to discharge and the output voltage remains constant until the input voltage is decreased enough to enable diode (B) to conduct. Therefore, the characteristic shown in figure a.3(*b*) describes the behaviour of the analogue element, which is a simulation of backlash, or hysteresis.

a.4 Continuous non-linear functions

An example of a 'continuous' non-linear function is an output from an element which is the square of the input and can be obtained using diodes. Diodes can be arranged such that they 'switch' resistors into a circuit for various values of input voltage and thus provide a varying value of resistance.

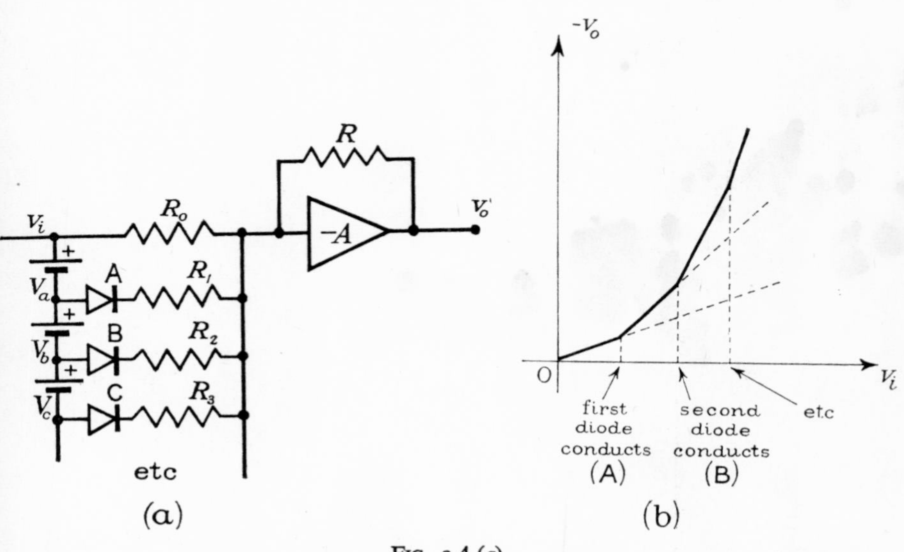

FIG. a.4 (*a*)

Figure a.4(*a*) shows an analogue element for a 'continuous' function. When the input voltage is increased the diodes A, B, C, etc., provide conducting paths in turn, adding a parallel resistor to the input resistor R_0. This increases the gain of the operational amplifier at each value of input voltage at which the diodes conduct, and produces the characteristic shown in figure a.4(*b*). The straight lines can approximate to a continuous curve, the form of which (square law, cubic, etc.) is adjusted by the bias voltages V_a, V_b, V_c, etc., and the values of resistors R_0, R_1, R_2, etc.

The bias supply is usually from a potentiometer chain connected to one supply voltage. The batteries have been included in figure a.4(*a*) for simplicity.

The diodes in sections a.1 to a.4 are assumed to have ideal characteristics.

INDEX

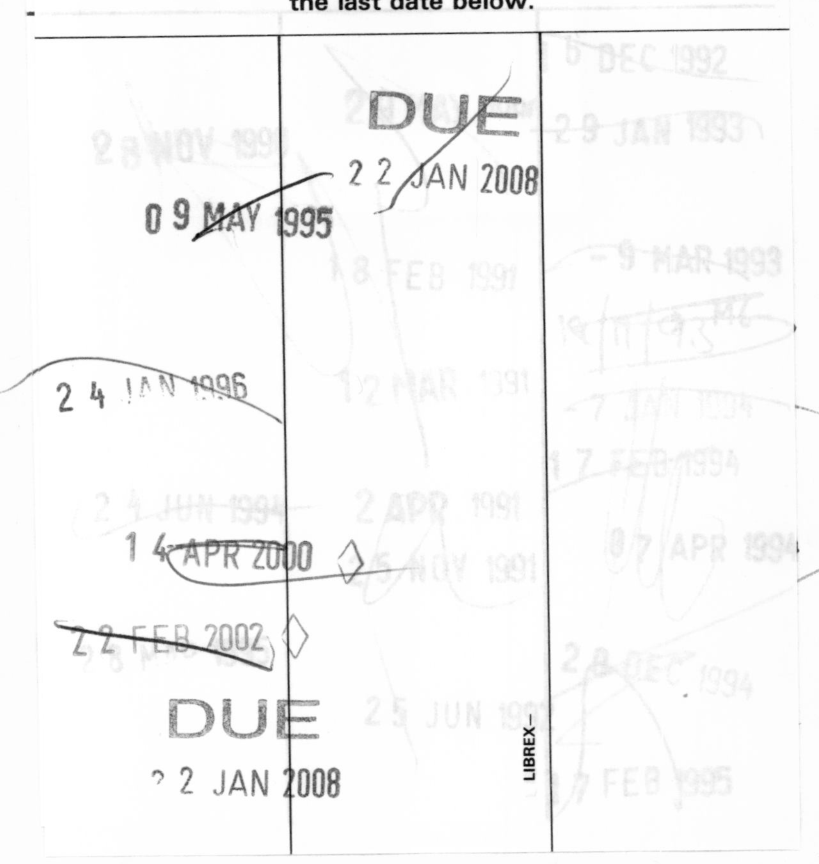
Books are to be returned on or before
the last date below.
DUE
22 JAN 2008
09 MAY 1995
24 JAN 1996
14 APR 2000
22 FEB 2002
DUE
22 JAN 2008
LIBREX

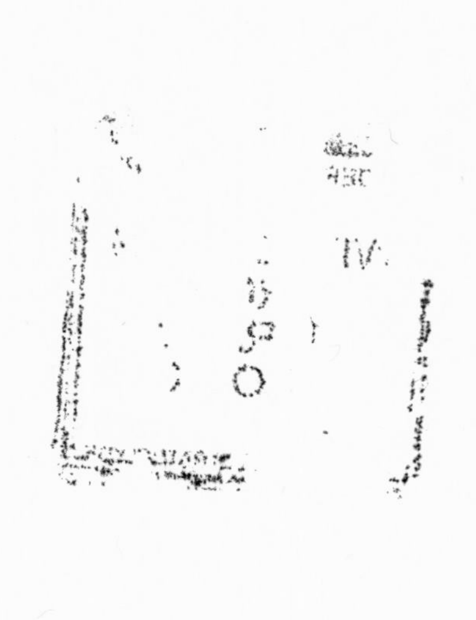